Understanding Meiosis and Mitosis

Understanding Meiosis and Mitosis

Stephanie Harris

MURPHY & MOORE
www.murphy-moorepublishing.com

Published by Murphy & Moore Publishing,
1 Rockefeller Plaza,
New York City, NY 10020, USA
www.murphy-moorepublishing.com

Understanding Meiosis and Mitosis
Stephanie Harris

International Standard Book Number: 978-1-63987-556-6 (Hardback)

This book contains information obtained from authentic and highly regarded sources. All chapters are published with permission under the Creative Commons Attribution Share Alike License or equivalent. A wide variety of references are listed. Permissions and sources are indicated; for detailed attributions, please refer to the permissions page. Reasonable efforts have been made to publish reliable data and information, but the authors, editors and publisher cannot assume any responsibility for the validity of all materials or the consequences of their use.

Trademark Notice: Registered trademark of products or corporate names are used only for explanation and identification without intent to infringe.

Cataloging-in-Publication Data

Understanding meiosis and mitosis / Stephanie Harris.
    p. cm.
Includes bibliographical references and index.
ISBN 978-1-63987-556-6
1. Meiosis. 2. Mitosis. 3. Cell division. 4. Cell cycle. 5. Biology. I. Harris, Stephanie.
QH605 .U53 2022
571.845--dc23

Table of Contents

Preface   VII

Chapter 1   Introduction   1
- Mitosis   1
- Meiosis   6
- Mitosis vs. Meiosis   57
- Genetic Variation in Mitosis and Meiosis   60

Chapter 2   Cell Cycle, Division and Differentiation   65
- Cell   65
- Cell Cycle   69
- G_0 Phase   70
- Interphase   77
- Apoptosis   88
- Cell Cycle Checkpoints   103
- Kinetochore   120
- Cell Growth   132
- Cell Division   132
- Cytokinesis   137
- Cell Differentiation   141

Chapter 3   Phases in Mitosis   157
- Prophase   158
- Prometaphase   160
- Metaphase   161
- Anaphase   163
- Telophase   164
- Factors that affect Mitosis   169

Chapter 4   Phases in Meiosis   172
- Meiosis 1   172
- Meiosis 2   185
- Difference between Meiosis 1 and Meiosis 2   187

Chapter 5   Genetic Recombination and Crossing Over   190
- Recombination   190
- Genetic Recombination   190
- Recombination in Mitosis and Meiosis   206
- Chromosomal Crossing Over   211
- Somatic or Mitotic Crossing Over   223
- Germinal or Meiotic Crossing Over   224

Permissions

Index

Preface

The purpose of this book is to help students understand the fundamental concepts of this discipline. It is designed to motivate students to learn and prosper. I am grateful for the support of my colleagues. I would also like to acknowledge the encouragement of my family.

Meiosis and mitosis are the processes of cell division that are studied in cell biology. Meiosis is a type of cell division that is used to produce gametes like sperm or egg cells. It is used by sexually reproducing organisms. This process includes two rounds of cell division that leads to the formation of four cells with one copy of each chromosome. Mitosis is the process in which chromosomes are replicated into two new nuclei. This results in cells that are genetically identical and which retain the same number of chromosomes. It is concerned with the transfer of parent cell's genome into two subsequent daughter cells. The processes of meiosis and mitosis differ in two aspects. These are recombination and the number of chromosomes. The topics included in this book are of utmost significance and bound to provide incredible insights to readers. Different approaches, evaluations, methodologies and studies related to this field have been included herein. Coherent flow of topics, student-friendly language and extensive use of examples make this book an invaluable source of knowledge.

A foreword for all the chapters is provided below:

Chapter – Introduction

The process of splitting up of a single cell to produce four cells containing half the original amount of genetic information is called meiosis. Mitosis is defined as the process in a cell cycle in which replicated chromosomes are separated into two new nuclei. This chapter has been carefully written to provide an easy understanding of meiosis and mitosis.

Chapter – Cell Cycle, Division and Differentiation

A series of events that takes place in a cell as it grows and divides is termed as a cell cycle. The process in which a parent cell is divided into two or more daughter cells is called cell division. Cell differentiation refers to the process in which the dividing cells alter their functional type. This chapter discusses cell cycle, cell division and cell differentiation in detail.

Chapter – Phases in Mitosis

The process of mitosis mainly consists of five major phases to undergo genetically identical cell division. It includes prophase, prometaphase, metaphase, anaphase and telophase. This chapter closely examines these phases of mitosis to provide an extensive understanding of the subject.

Chapter – Phases in Meiosis

The phases of meiosis can be divided into meiosis I and meiosis II which are further divided into karyokinesis I and cytokinesis I and karyokinesis II and cytokinesis II respectively. It involves prophase 1, metaphase 1, anaphase 1 and telophase 1. The topics elaborated in this chapter will help in gaining a better perspective of the different phases of meiosis.

Chapter – Genetic Recombination and Crossing Over

Genetic recombination refers to the exchange of genetic material between multiple chromosomes or between different regions of the same chromosome for production of offspring with traits of parents. The exchange of chromosome segments between non-sister chromatids during the production of gametes is termed as crossing over. All the aspects related to genetic recombination and crossing over have been carefully analyzed in this chapter.

Stephanie Harris

Introduction

The process of splitting up of a single cell to produce four cells containing half the original amount of genetic information is called meiosis. Mitosis is defined as the process in a cell cycle in which replicated chromosomes are separated into two new nuclei. This chapter has been carefully written to provide an easy understanding of meiosis and mitosis.

MITOSIS

Mitosis constitutes a comparatively small portion of a complete cell cycle but it is one of the imperative parts of the cell cycle. German Physician and cell biologist "Walther Flemming" coined the term "mitosis" in the year 1882. He explained the process of how cells split and separate their chromosome.

The process of cell division that results in the formation of two new daughter cells is termed as Mitosis. The newly formed daughter cells are genetically identical to the parent cell and to each other. It plays a crucial role in a living organism's life cycle. However, the level of significance may vary depending on the type of organism (multicellular or single-celled).

In unicellular organisms such as bacteria, mitosis helps in asexual reproduction as it produces an identical copy of the parent cell. Another example of the Eukaryotic unicellular organism is "Amoeba." An amoeba uses cell division for the production of new individuals. In the case of multicellular organisms, mitosis helps in growth and repair by producing more number of identical cells. For example plants, animals depend on cell division for their growth by addition of new cells. It is also used for repairing the injured tissues or replacing the worn-out tissue by regenerating new cells.

Mitosis refers to the splitting of chromosomes in the eukaryotic cells during the cell division process. The parent cell divides into two daughter cells that are identical to the parent cell during the process of cell division. During the mitosis process, the cell's nucleus along with the chromosome is divided to form two new daughter cell nuclei. The daughter nuclei inherit the same number of chromosomes as that of the parent nucleus.

Importance of Mitosis in Living Process

- Genetic stability: Mitosis helps in the splitting of chromosomes during cell division and generates two new daughter cells. Therefore the chromosomes form from the parent chromosomes by copying the exact DNA. Therefore, the daughter cells formed as genetically uniform and identical to the parent as well as to each other. Thus mitosis helps in preserving and maintaining the genetic stability of a particular population.

- Growth: Mitosis help in increasing the number of cells in a living organism thereby playing a significant role in the growth of a living organism.

- Replacement and regeneration of new cells: Regeneration and replacement of worn-out and damaged tissues is a very important function of mitosis in living organisms. Mitosis helps in the production of identical copies of cells and thus helps in repairing the damaged tissue or replacing the worn-out cells. But the degree of regeneration and replacement in multicellular organisms vary from one another. For example, mitosis process is used in order to regrowth the legs of newts and crustaceans. However, the degree of regrowth may vary.

- Asexual reproduction: Mitosis is used in the production of genetically similar offspring. For example budding of hydra and yeast, binary fission in amoeba, etc.

Significance of Mitosis

Mitosis plays an important role in the life of living organisms in various ways as given below:

- After fusion of male and female gametes zygote is formed. Mitosis is responsible for development of a zygote into adult organism.

- Mitosis is essential for normal growth and development of living organisms. It gives a definite shape to a specific organism.

- In plants, mitosis leads to formation of new parts, viz., roots, leaves, stems and branches. It also helps in repairing of damaged parts.

- In case of vegetatively propagated crops like sugarcane, sweet potato, potato, etc., mitosis helps in asexual propagation. Mitosis leads to production of identical progeny in such crops.

- Mitosis is useful in maintaining the purity of types because it leads to production of identical daughter cells and does not allow segregation and recombination to occur.

- In animals, it helps in continuous replacement of old tissues with new ones, such as gut epithelium and blood cells.

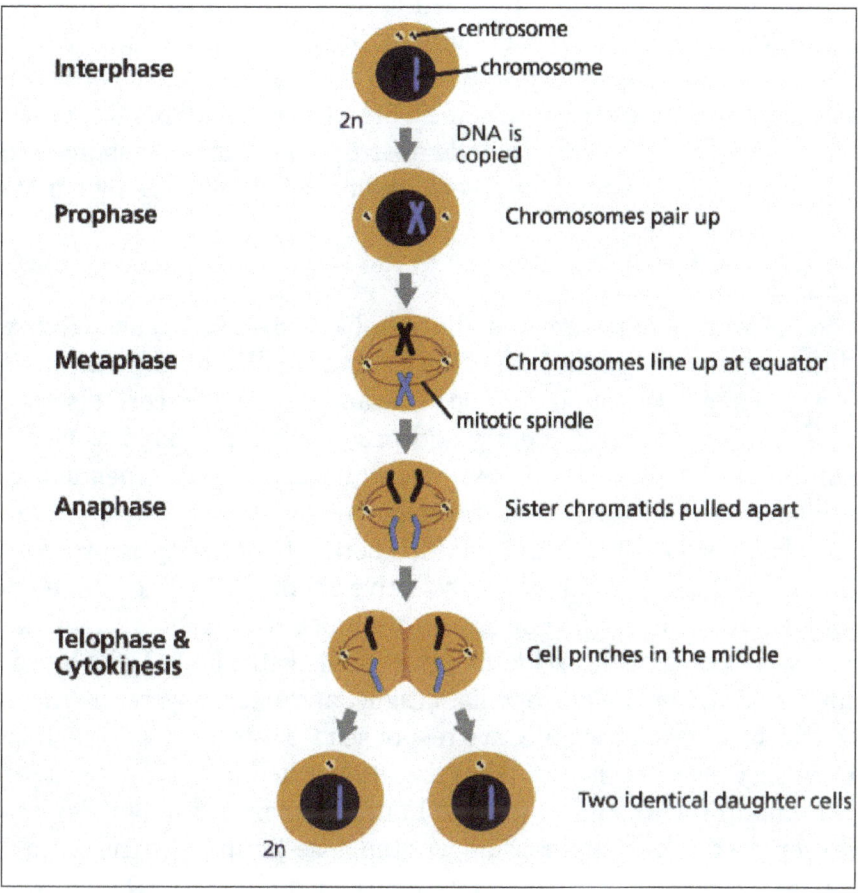

Mitotic Inhibitors

Mitotic inhibitors are drugs derived from natural plant sources. They inhibit cell division or mitosis, where a single cell divides into two genetically identical daughter cells. Mitotic inhibitors bind to tubulin and inhibit its polymerization into microtubules. Microtubules are structures responsible for pulling the cell apart when it divides. Mitotic inhibitors affect cancer cells more than normal cells because cancer cells divide (mitotic cell division) more rapidly therefore are more susceptible to mitotic inhibition. Different mitotic inhibitors are used to treat particular types of cancers, such as leukemia, lymphoma, breast cancer, lung cancer and other types of cancers.

Aurora Kinase Inhibitors

The three mammalian aurora kinases (A, B, and C) are centrosome-associated serine/threonine kinases involved in spindle assembly, centrosome development, chromosome segregation, and cytokinesis during mitosis and meiosis. Overexpression of aurora A and aurora B has been documented in several common cancers including

NSCLC. Aurora B overexpression has been associated with reduced overall survival in NSCLC.3. Several different aurora kinase inhibitors, affecting A, B, or both, plus or minus other kinases are already in phase I and II trials. Both oral and intravenously administered agents are being explored, using a broad range of administration schedules. Dose-limiting toxicities commonly include neutropenia. Clinical responses in solid tumors have been modest to date. Biomarkers predictive for benefit remain undiscovered.

Polo-like Kinase Inhibitors

Plk1 is the best characterized member of the polo-like kinase family and is overexpressed in many different cancers, particularly NSCLC and small cell lung cancer. Higher expression is associated with both increased stage and grade of cancer. It is a serine/threonine kinase required for mitotic entry and progression, actively involved in centrosome maturation and spindle formation, undergoing cell-cycle dependent changes in both intracellular spatial distribution and enzymatic activity. PLk1 inhibition is known to lead to the expression of PLK2 and 3, which both inhibit cell cycle progression. Data were presented on BI2536, a potent and selective adenosine triphosphate-competitive inhibitor of Plk1. A phase II trial comparing two dose schedules (200 mg day 1 versus 40–50 mg days 1–3 on a 21-day schedule) in patients with NSCLC who had failed first-line therapy was described. Neutropenia, fatigue, and nausea were the most common severe side effects. Overall, the objective response rate was 4.2% with no apparent differences in efficacy between the two administration schedules. BI6727 is a second-generation Plk1 inhibitor with a half-life >100 hours. A phase I trial of a day 1 infusion of BI2536 was reported. Myelosuppression was the dose-limiting toxicity with the monotherapy maximum tolerated dose defined as 400 mg and the RP2D defined as 320 mg every 21 days. Two partial responses were seen among the 50 patients within the phase I monotherapy study (1 × urothelial and 1 × ovarian cancer). A three-arm randomized phase II study in the second/third-line setting for advanced NSCLC of pemetrexed versus 300 mg BI6727 versus the combination of the two (with the BI6727 dose confirmed in a safety lead-in phase in the same study) is planned.

CENP-E Inhibitors

The mitotic checkpoint (or spindle assembly checkpoint) acts to prevent chromosomal missegregation and subsequent aneuploidy. The centromere-linked, kinesin-like motor protein CENP-E (Kinesin 7) contributes to the mitotic checkpoint. It is hypothesized that in tissues (e.g., malignant tissues) with elevated basal levels of chromosomal instability, additional instability from CENP-E dysfunction/inhibition is enough to trigger cell death from mitotic catastrophe and loss of all copies of specific essential genes. Updated data were presented on GSK923295, a potent and selective small molecule inhibitor of CENP-E. Preclinically, amplification of the cMyc oncogene was associated with increased sensitivity to GSK923295, with the ratio of sensitive to resistant cell lines being 11 to 0 in cMyc amplified lines and 25 to 15 in nonamplified lines. A phase I dose escalation study of intravenously administered GSK923295 on days 1, 8, and 15 of a 28-day cycle is ongoing with doses up to 190 mg/m^2 being tolerable to date. A phase II study of GSK923295 in NSCLC is planned.

Chk1 Inhibitors

Chk1 is a serine/threonine DNA damage response kinase associated with checkpoint pathways at the S and G2/M phases of mitosis. The G1 checkpoint is often compromised in cancer cells, potentially elevating the importance of S and G2/M checkpoints. By inhibiting Chk1, cancer cells may be more prone to proceed through the checkpoint after DNA damage leading to mitotic catastrophe. Data were presented on AZD7762, a potent and selective Chk inhibitor. Preclinically, it potentiates the effects of gemcitabine in H460 xenograft models, with evidence of increased effectiveness in p53 null cells (i.e., in those with already abrogated G1 checkpoints). A phase I study of intravenous AZD7762 in combination with either weekly gemcitabine or weekly irinotecan is ongoing.

Kinesin Spindle Protein Inhibitors

Kinesin spindle protein (KSP) is a mitotic kinesin, involved in spindle pole separation and the maintenance of spindle integrity. Several different companies are developing drugs that inhibit KSP, including GSK, AstraZeneca, Merck, and Array. To date, KSP inhibitors seem to be associated with neutropenia as the major dose-limiting toxicity. Monotherapy efficacy signals for GSKs Ispinesib have been noted in the second or third-line treatment of advanced breast cancer, with a reported 9% response rate. The potential mechanisms underlying clinical responses to KSP and other key mitotic inhibitors that induce cell cycle arrest. It was reported that Array-520 shows significantly more activity in models of hematological malignancies than in models of solid tumors. Notable, prolonged in vitro exposure to Array-520 produces an apoptotic peak earlier in myeloma cell lines than in colorectal cell lines.

A hypothesis explaining these results was proposed. Specifically, during cell cycle arrest, protein synthesis is inhibited but protein degradation may remain intact. Consequently, the time the cell remains in arrest, before achieving either mitotic slippage or damage repair and mitotic progression, will determine whether supplies of certain survival proteins will be degraded to levels below which the cell is committed to apoptotic cell death. An individual cell's basal levels and the half-life of these survival proteins will provide further relevant variables, in addition to the duration of arrest that will determine outcome. Data were presented showing that one major difference between myeloma and colorectal cell lines was the duration of Mcl-1 expression after cell cycle arrest, the more rapid disappearance in the myeloma cells correlating with the early peak of apoptosis. High levels of mcl-1 are known to be associated with a range of different malignancies including SCLC and NSCLC; however, the full gamut of factors that may be important in determining sensitivity to mitotic inhibitors through survival protein dependence mechanisms is only just beginning to be explored. On the basis of these preclinical results, more dose intensive schedules to prolong the period of arrest are being explored in phase I studies of Array 520 in both solid tumors and hematological malignancies.

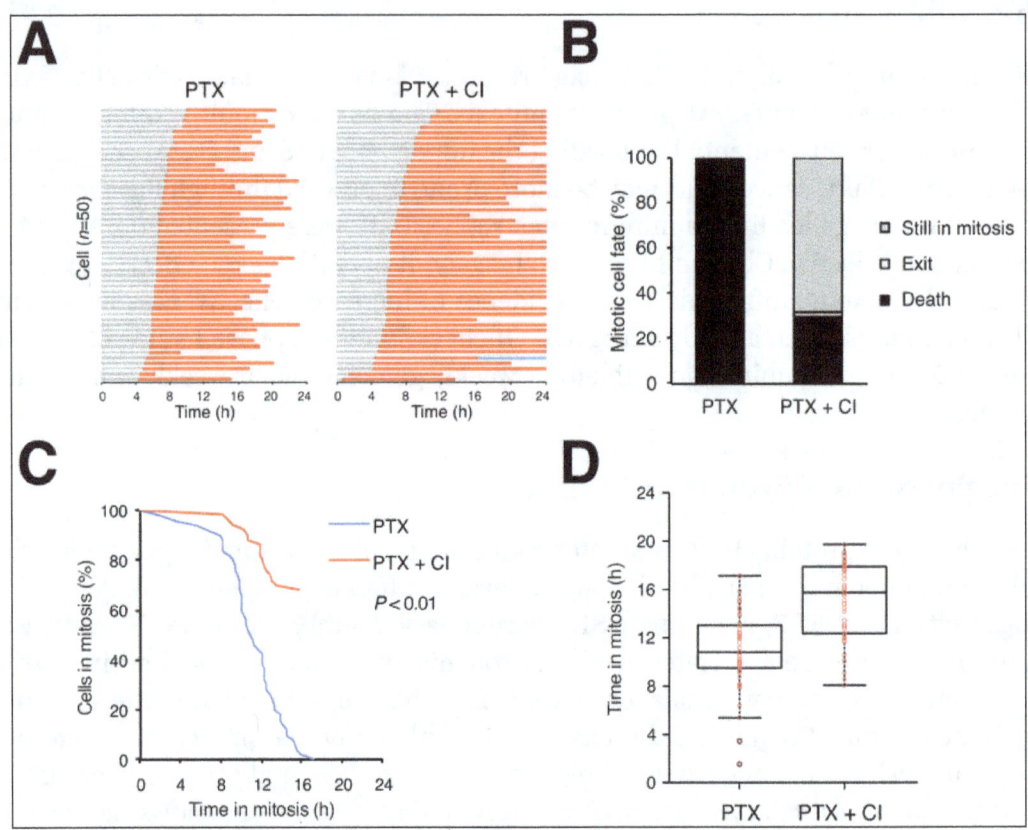

Half-life of key survival proteins during cell cycle arrest determines the onset of apoptosis.

MEIOSIS

Meiosis is a type of cell division that reduces the number of chromosomes in the parent cell by half and produces four gamete cells. This process is required to produce egg and sperm cells for sexual reproduction. During reproduction, when the sperm and egg unite to form a single cell, the number of chromosomes is restored in the offspring.

Meiosis begins with a parent cell that is diploid, meaning it has two copies of each chromosome. The parent cell undergoes one round of DNA replication followed by two separate cycles of nuclear division. The process results in four daughter cells that are haploid, which means they contain half the number of chromosomes of the diploid parent cell.

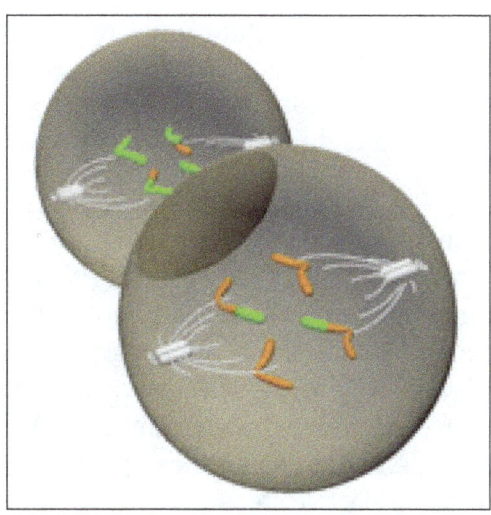

Meiosis has both similarities to and differences from mitosis, which is a cell division process in which a parent cell produces two identical daughter cells. Meiosis begins following one round of DNA replication in cells in the male or female sex organs. The process is split into meiosis I and meiosis II, and both meiotic divisions have multiple phases. Meiosis I is a type of cell division unique to germ cells, while meiosis II is similar to mitosis.

Meiosis I, the first meiotic division, begins with prophase I. During prophase I, the complex of DNA and protein known as chromatin condenses to form chromosomes. The pairs of replicated chromosomes are known as sister chromatids, and they remain joined at a central point called the centromere. A large structure called the meiotic spindle also forms from long proteins called microtubules on each side, or pole, of the cell. Between prophase I and metaphase I, the pairs of homologous chromosome form tetrads. Within the tetrad, any pair of chromatid arms can overlap and fuse in a process called crossing-over or recombination. Recombination is a process that breaks, recombines and rejoins sections of DNA to produce new combinations of genes. In metaphase I, the homologous pairs of chromosomes align on either side of the equatorial plate. Then, in anaphase I, the spindle fibers contract and pull the homologous pairs, each with two chromatids, away from each other and toward each pole of the cell. During telophase I, the chromosomes are enclosed in nuclei. The cell now undergoes a process called cytokinesis that divides the cytoplasm of the original cell into two daughter cells. Each daughter cell is haploid and has only one set of chromosomes, or half the total number of chromosomes of the original cell.

Meiosis II is a mitotic division of each of the haploid cells produced in meiosis I. During prophase II, the chromosomes condense, and a new set of spindle fibers forms. The chromosomes begin moving toward the equator of the cell. During metaphase II, the centromeres of the paired chromatids align along the equatorial plate in both cells. Then in anaphase II, the chromosomes separate at the centromeres.

The spindle fibers pull the separated chromosomes toward each pole of the cell. Finally, during telophase II, the chromosomes are enclosed in nuclear membranes. Cytokinesis follows, dividing the cytoplasm of the two cells. At the conclusion of meiosis, there are four haploid daughter cells that go on to develop into either sperm or egg cells.

The origin of meiosis through gradual steps is among the most intriguing evolutionary enigmas. Meiosis is one of the 'major innovations' of eukaryotes that evolved before their subsequent radiation over 1 billion years ago. Extant eukaryotes share a set of genes specifically associated with meiosis, implying that it evolved only once before their last common ancestor. Identifying the selective scenario that led to its early evolution is difficult, but clues can be obtained by determining: (i) Which mitotic cellular processes were reused in meiosis (e.g. DNA repair through homologous recombination and possibly reduction), (ii) Which selective steps were involved in the assembly of the full cellular process, and (iii) Why different forms of meiosis were perhaps less successful.

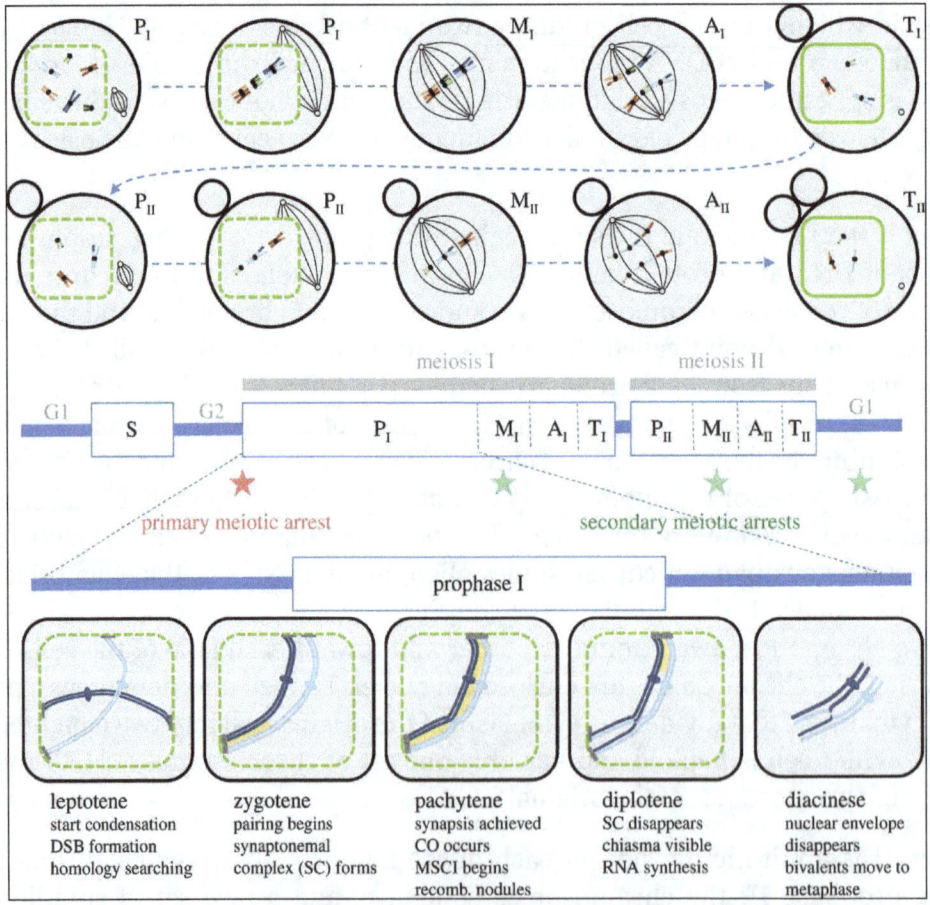

The different steps in standard meiosis.

A form of reductional cell division (aka 'proto-meiosis') probably evolved in early asexual unicellular eukaryotes. Two scenarios for this have been proposed. The first is that diploidy accidentally occurred by replication of the nuclear genome without subsequent cell division (endoreplication), and that returning to haploidy was selected for to correct this. Because either haploidy or higher ploidy levels may be favoured in different ecological situations, a variant of this scenario is that a proto-meiosis–endoreplication cycle evolved to switch between ploidy levels. The resulting life cycle may have resembled modern 'parasexual' fungi in which diploid cells lose chromosomes in subsequent mitotic divisions, leading to haploidy via aneuploid intermediates. Many other modern eukaryotes also increase and decrease their ploidy somatically, depending on growth stage or specific environmental stimuli. The second scenario is that proto-meiosis evolved in response to the fusion of two haploid cells (syngamy), as in standard modern eukaryotic sexual life cycles. Syngamy may have been favoured because it allows recessive deleterious mutations to be masked in diploids. A difficulty with this idea is that such masking may not be sufficient to favour diploidy in asexuals. In a variant of this scenario, early syngamy evolved as a result of 'manipulation' by selfish elements (plasmids, transposons) to promote their horizontal transmission. In support of this view, mating-type switching (which can allow syngamy in haploid colonies) has evolved multiple times in yeasts and involves domesticated mobile genetic elements.

The top panel illustrates the different phases of a typical female meiosis for each of the two meiotic divisions: Prophase (P, with early and late prophase distinguished), metaphase (M), anaphase (A) and telophase (T). The nuclear membrane is indicated by the green contour (dashed when it starts fragmenting). The small black circles represent microtubule organizing centres and the black lines represent microtubules of the meiotic spindle. First and second polar bodies are shown as grey circles next to the oocyte (chromosomes inside the polar bodies are not shown). Homologous chromosomes are represented with the same colour with slightly different shades (e.g. orange and light orange). Homologues pair and segregate in meiosis I, then sister chromatids segregate in meiosis II. The middle panel shows the meiotic cell cycle. The timing of the primary meiotic arrest is indicated by a red star, while the timing of the most common secondary arrests in different organisms is indicated by green stars. The lower panel indicates the important steps (DSB formation, crossing overs) occurring during prophase I. The synaptonemal complex is shown in yellow. Chromatin condenses in chromosomes throughout prophase I (only one pair of homologues is illustrated). In most species, telomeres attach to the nuclear envelope. The attachment plate is indicated by a grey bar. MSCI, meiotic sex chromosome inactivation.

Meiosis requires the correct segregation of homologues, which is achieved by homologue pairing at the beginning of prophase I. This homology search is mediated by the active formation of numerous DNA double-strand breaks (DSBs) followed by chiasmata formation, but less well-known mechanisms of recombination-independent pairing also exist. Non-homologous centromere coupling is also often observed at this stage, but the functional and evolutionary significance of this coupling is elusive. In many species, chromosome pairing is further strengthened by 'synapsis', which is the formation of a protein structure known as the synaptonemal complex and the pairing of homologous centromeres. Chiasmata are then resolved as either crossovers (hereafter 'COs') resulting in the exchange of large chromatid segments, or non-crossovers (NCOs), where both situations cause gene-conversion events. The synaptonemal complex then disappears, and homologues remain tethered at CO positions and centromeres. The precise function of the synaptonemal complex is not entirely understood ; one possibility is that it may serve to stabilize homologues during CO maturation. Some pairing mechanism must be advantageous to ensure proper segregation of homologues, but the origins and selective advantage of extensive pairing, synapsis, gene conversion and recombination remain poorly understood.

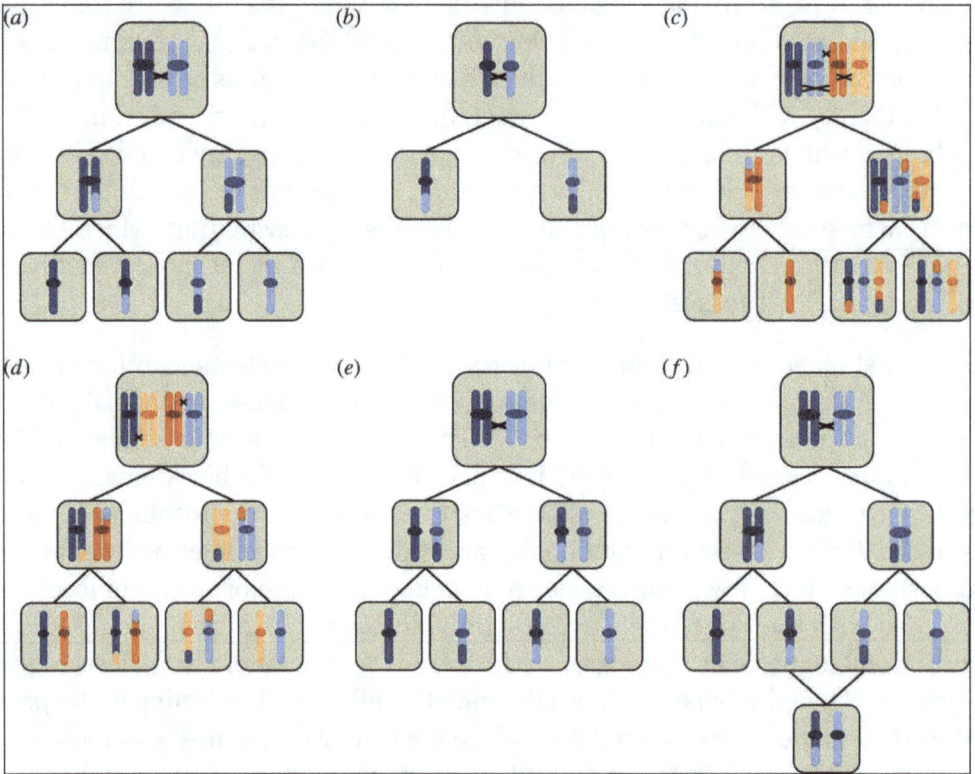

Most evidence suggests that homologous recombination evolved long before meiosis, as it occurs in all domains of life and involves proteins that share strong homology. One hypothesis is that meiotic pairing and extensive homologous recombination in meiosis evolved to avoid the burden and consequences of non-allelic ectopic recombination in the large genomes of early eukaryotes, which presumably had many repetitive sequences. Such sequences may have been related to the spread of retrotransposons in early eukaryotes, of which many types are very ancient in eukaryotes, but absent in bacteria and archaea. A second possibility is that recombination arose by the spread of self-promoting genetic elements exploiting the machinery of DNA repair and associated gene conversion. Another hypothesis is that pairing and recombination initially arose as a way to repair mutational damage caused by increased oxidative stress due to rising atmospheric oxygen or endosymbiosis. This scenario presupposes that DNA maintenance is inefficient in the absence of meiosis; however, prokaryotes (including archaea) have efficient repair mechanisms that involve recombination, but not meiosis. In addition, this scenario does not fit well with the observation that a large number of DSBs are actively generated at the onset of meiosis.

Meiosis and some of its modifications. (*a*) Regular meiosis. Following DNA replication, homologous chromosomes are separated in the first meiotic division, whereas sister chromatids are separated in the second division. COs result in chromosomes in the final meiotic products that carry genetic material from both homologous chromosomes. (*b*) Hypothetical 'one-step' meiosis, in which DNA replication before entering meiosis is suppressed and, therefore, only a single meiotic division is required. (*c*) Multivalent formation in a neo-tetraploid. Blue and orange chromosome pairs are assumed to be identical or very similar so that pairing can occur. Chiasmata of one chromosome with three other chromosomes leads to mis-segregation. (*d*) Bivalent formation in a tetraploid with exactly one CO per chromosome. Chromosomes may pair randomly (leading to polysomic inheritance), but segregation proceeds normally. (*e*) Inverted meiosis, in which sister chromatids are separated in the first division and homologous chromosomes in the second division. Note that although centromeres are shown here for clarity, all described species consistently using inverted meiosis are holokinetic (no centromeres). (*f*) Central fusion automixis, a mechanism of producing diploid eggs that can then develop parthenogenetically without fertilization. As a consequence of COs, heterozygosity may be lost with this mechanism in regions distal to the centromere.

A particular feature of meiosis is that it starts with chromosome doubling before meiosis occurs. For ploidy reduction, the initial steps appear superfluous. A simpler single-step cell division, without the initial DNA replication phase, could in principle achieve ploidy reduction. Recombination may not be a crucial difference between one- and two-step meiosis, as both can involve COs, even if with one CO, the two meiotic products carry recombinant chromosomes in one-step meiosis, whereas only two out of four are recombinant in two-step meiosis. Three hypotheses have been proposed to account for two-step meiosis. The first postulates that two-step meiosis better protects against particular selfish genetic elements (SGEs) that increase their transmission frequencies by sabotaging the meiotic products in which they do not end up (known as 'sister killers', distinct from the 'sperm killers'). In a two-step meiosis, there is uncertainty as to whether the reductional division is meiosis I or II, meaning that the sabotage mechanism has a much reduced efficacy.

Microsporidia and red algae show specific modifications to meiosis that increase such uncertainty even more. However, such sister killers are hypothetical, and theoretical studies based on assumptions about how different killers might act suggest that this mechanism does not inevitably promote the development of a two-step meiosis. The second hypothesis is that sexual species with one-step meiosis would be vulnerable to invasion by asexual mutants, and have thus gone disproportionally extinct in the past. Contrary to one-step meiosis, most automictic modifications of two-step meiosis involve a loss of heterozygosity with each generation which would cause expression of recessive and partially recessive deleterious mutational effects, and reduce the fitness of newly emerging asexual mutants. Finally, a third hypothesis posits that a one-step meiosis is more complex and thus less likely to evolve than a two-step meiosis. Mitotic and meiotic cell cycles start similarly with DNA replication in response to increasing cyclin-dependent kinase (CDK) activity. Two-step meiosis can be achieved simply by modulating CDK activity at the end of a cell cycle to add a second division event. By contrast, a one-step meiosis would require extensive modification of the mitotic cycle. Despite earlier suggestions of its presence in some basal eukaryotes (protists), there are presently no firm indications that one-step meioses exist in nature, although inverted meiosis is genetically similar to mitosis followed by single-step meiosis.

Secondary Modifications of Meiosis

Meiosis is remarkably conserved across eukaryotes. Nevertheless, in many species, variants exist that may offer insights into the evolutionary origins and mechanistic constraints of meiosis.

Meiosis and Polyploidy

Polyploidy is surprisingly common in eukaryotes given the considerable problems it poses to meiosis. In diploids, homologous chromosomes recognize each other and align to form bivalents during Prophase I, but when there are three or more chromosomes with sufficient homology, these chromosomes may all align to varying degrees, forming multivalents. This can occur when all chromosome sets originate from the same species (autopolyploidy), and also when polyploidy is a result of hybridization (allopolyploidy). Multivalent formation is often associated with mis-segregation of chromosomes as well as chromosomal rearrangements arising from recombination within multivalents, leading to reduced fertility and low-fitness offspring. These problems may be compounded in allopolyploids because recombination homogenises partially differentiated chromosomes, thereby further increasing the likelihood that they will pair (the 'polyploid ratchet').

The existence of successful polyploid species and lineages indicates that natural selection can often promote transitions from multi- to bivalents that will then segregate as in diploids. However, how such transitions are achieved at the molecular level remains a mystery. Part of the answer seems to be a reduction in the number of COs, since multivalents can only form with at least two COs per chromosome. This mechanism seems particularly important in autopolyploids and may be achieved through increasing CO interference. Several candidate genes that may effect such modifications have been identified in the autotetraploid *Arabidopsis arenosa*. In allopolyploids, there is evidence for genes that have been selected to strengthen the preferential pairing of homologous (i.e. of the same origin, rather than 'homeologous') chromosomes, including *ph1* in hexaploid wheat. This preferential pairing can also be achieved through reducing CO numbers, but specifically those between homeologues; this could indirectly produce an increase in CO numbers and hence recombination rates between homologues. Intriguingly, because most extant organisms have a history of polyploidy, many features of 'standard' meiosis such as CO interference may have been shaped by the problems involved in multivalent segregation.

Polyploidy with odd numbers of chromosome sets poses an even greater problem because aneuploid gametes are generally produced (e.g.). However, there are some plant species where solutions to even this problem have evolved, and where odd-number polyploidy appears to persist in a stable manner. In these species, the problem of unequal segregation during meiosis is solved through exclusion of univalents in one sex but inclusion in the other, leading, for example, to haploid sperm and tetraploid eggs in pentaploid dog roses.

Inverted Meiosis

In normal meiosis, homologous chromosomes are separated during meiotic division I, whereas sister chromatids are separated during meiosis II. Why meiosis generally follows this order is unknown, but interestingly, in some species meiosis takes place in the reverse order, including some flowering plants, mites, true bugs and mealybugs. All species with this 'inverted' meiosis described to date seem to have holocentric chromosomes (i.e. the kinetochores are assembled along the entire chromosome, rather than at localized centromeres). Inverted meiosis is viewed as a possible solution to specific problems of kinetochore geometry in such meiosis. Yet, intriguing as they are, these systems provide little insight into why inverted meiosis is absent or very rare in monocentric species.

It is conceivable that a reverse order of divisions would make meiosis more vulnerable to exploitation by meiotic drive or sister killer SGEs, but to the best of our knowledge, there is currently neither theoretical nor empirical support for this idea. Another possibility is that meiosis I tends to be reductional because it allows for DSB repair by sister chromatid exchange in arrested female meiosis. Alternatively, the order of meiotic divisions could merely be a 'frozen accident', i.e. a solution that has been arrived at a long time ago by chance, and that reversal is difficult (at least with monocentric chromosomes).The careful genotyping of eggs (or embryos) and polar bodies at many markers indicated that surprisingly often, chromosomes followed an 'inverted meiosis' pattern of segregation, even though this led to aneuploidies in approx. 23% of cases.

Meiosis Modifications and Loss of Sex

Many organisms have abandoned canonical sexual reproduction, reproducing asexually by suppressing or modifying meiosis and producing diploid eggs that can develop without fertilization. This raises two connected mysteries: why are some types of modifications much more frequent than others, and how can mitotic (or mitosis-like) asexual reproduction ('apomixis' or 'clonal parthenogenesis' in animals, 'mitotic apomixis' in plants) evolve from meiosis? Examples of meiosis-derived modes of asexual reproduction include chromosome doubling prior to meiosis ('endomitosis' or 'pre-meiotic doubling'), fusion of two of the four products of a single meiosis ('automixis' in animals, 'within-tetrad mating' in fungi), and suppression of one of the two meiotic divisions.

Two particularly common modes of asexuality are the suppression of meiosis I, and automixis involving fusing meiotic products that were separated during meiosis I ('central fusion',). Both are genetically equivalent and lead to reduced heterozygosity when there is recombination between a locus and the centromere of the chromosome on which it is located. Most other forms of meiosis-derived asexual reproduction lead to a much stronger reduction in offspring heterozygosity, and it has been hypothesized that the reduced fitness of homozygous progeny explains the rarity of these other forms. Indirect support comes from the observation that species with regular asexual reproduction usually do so by central fusion or suppression of meiosis I, often accompanied by very low levels of recombination, thus maintaining heterozygosity. By contrast, species that only rarely reproduce asexually show a wider variety of asexual modes and higher levels of recombination. Nonetheless, this hypothesis cannot explain some observations, for instance, the rarity of pre-meiotic doubling with sister-chromosome pairing, which would also efficiently maintain heterozygosity. Perhaps, evolving a mechanism that ensures exclusive sister pairing (i.e. the complete absence of non-sister pairing) is difficult, though it seems to occur in some lizard species. In addition, such a system would make it difficult to repair DSBs occurring before doubling (as both sister chromatids would have the same DSBs).

However, unless meiosis can be entirely bypassed (e.g. as with vegetative reproduction), secondary asexuality is likely to evolve via modification of meiosis, keeping much of the cell signalling and machinery intact. Indeed, detailed cytological and genetic investigations in several asexual species thought to reproduce clonally by mitotic apomixis have uncovered remnants of meiosis. In *Daphnia*, meiosis I is aborted mid-way and a normal meiosis II follows. Hence, clonality in *Daphnia* is meiotically derived. This should lead to loss of heterozygosity in centromere-distal regions, but if recombination is fully suppressed the genetic outcome resembles mitosis. Importantly, this suggests a possible stepwise route to evolution of mitosis-like asexuality. Rare automixis (spontaneous development of unfertilized eggs) occurs in many species. If this becomes more common, forms of automixis maintaining heterozygosity in centromere regions might be selectively favoured and recombination suppressed, eventually leading to meiosis-derived asexuality with the same genetic consequences as mitosis. Indeed, in *Arabidopsis*, meiosis can be transformed to genetically resemble mitosis, but modification of several genes is needed to achieve this. In angiosperms, there is also the difficulty to overcome the absence of endosperm fertilization to achieve proper seed development, which further stresses that meiosis-derived asexuality is unlikely to evolve in a single step. To fully understand the evolutionary maintenance of sex, we may therefore need

to understand the selection pressures acting in the intermediate stages, which probably involve loss of heterozygosity, and thus inbreeding depression. In many cases, the
initial evolution of asexuality may thus resemble the evolution of self-fertilization, and
several traits may pre-exist (such as low recombination rates) that make the successful
transition to asexuality more likely in some taxa.

Meiosis Punctuates Life Cycles

Meiosis is a key step in sexual life cycles, as well as some asexual life cycles derived from
sexual ancestors. In multicellular eukaryotes, where meiosis is tightly associated with
reproduction (unlike in many protists), meiosis is also a cellular and genetic bottleneck
at the critical transition between the diploid and the haploid phases.

Meiosis Timing and Arrest

In early haploid eukaryotes, meiosis probably quickly followed endomitosis or syngamy. Today, multicellular eukaryotes exhibit a variety of life cycles in which the haploid or diploid phase may predominate. The duration of the different phases was perhaps initially controlled, in part, by the timing of meiosis—for instance, a multicellular,
extended diploid phase likely evolved by postponing meiosis. However, in metazo-
ans, life cycles are mostly determined by the extent of somatic development within
each phase rather than by the timing of meiosis, which can be halted or postponed. In
animals, where haploid mitosis is suppressed, syngamy immediately follows meiosis.
Furthermore, specific cells are 'destined' at an early stage to eventually undergo meiosis
(aka germline), whereas this cell fate is determined much later in fungi, plants and some
algae.

The timing of meiosis in the germline of animals has been intensively investigated.
Whereas male meiosis occurs continuously, female meiosis usually stops twice. These
'meiotic arrests' are under the control of various factors that are not completely identified across animals. Arrest 1 occurs in prophase I during early development and
can last years until sexual maturity. The timing of arrest 2 is more variable (ranging
from metaphase I in many invertebrates, to metaphase II in vertebrates and G1 phase
after meiosis II in some echinoderms), and may have evolved to prevent the risk of
premature parthenogenetic cleavage of oocytes or inappropriate DNA replication before fertilization ; this is supported by the fact that this arrest is usually released by
fertilization. However, the evolutionary significance of its precise timing in diverse
groups is not well understood. Three ideas have been put forward to explain arrest 1.
First, its occurrence at prophase I may allow the repair of accidental DSBs by sister
chromatid exchange during long periods between arrests 1 and 2. Second, if arrest
1 was to occur during an earlier mitotic division within the germline, this might decrease the variance in the number of deleterious mutations among gametes within individuals, which may be detrimental if some defective gametes or early embryos can
be eliminated and replaced during reproduction. Third, it may be easier to prevent

uncontrolled proliferation in a non-dividing meiotic oocyte, as once the cell starts the meiotic cell division, it cannot engage in further mitotic divisions. Arrest 1 may thus have evolved to control (and minimize) the number of possibly wasteful and mutagenic mitotic divisions in the female germline. Similar meiotic arrests in plants are unknown. Plants seem to completely lack strict mechanisms to arrest the meiotic cell division. Contrary to animals and fungi that may arrest the cell cycle and abort meiosis once DSBs are not repaired, plants will progress through meiosis irrespective of such major defects.

Meiosis and Epigenetic Reset

Meiosis and syngamy represent critical transitions between haploid and diploid phases in each generation. It has been suggested that a primary function of meiosis is to allow for epigenetic resetting in eukaryotes. For instance, metazoan development is under the control of many epigenetic changes (cytosine methylation and chromatin marks) that are irreversibly maintained throughout life and must be reset twice each generation (at the $n \rightarrow 2n$ and $2n \rightarrow n$ transitions). This ensures proper development, the acquisition of parent-specific imprints, and may allow for mechanisms limiting the maximal number of possible successive mitoses ('Hayflick limit', reducing tumour development). Some loci escape these resets, which can lead to transgenerational epigenetic inheritance. This occurs much less frequently in animals than in plants (e.g. in *Arabidopsis*, demethylation is largely restricted to asymmetric CHH methylation sites, and contrary to mouse, does not occur on most symmetric CG and CHG methylation sites). Although the $2n \rightarrow n$ resetting occurs at or very close to meiosis in some cases (in female meiosis in animals), its timing may not be strictly tied to meiosis. For instance, it occurs pre-meiotically in the male germ line of animals or post-meiotically in male plant gametophytes.

The evolutionary significance of these timing differences are poorly understood. Meiosis may simply not be the optimal time for epigenetic resetting. Many epigenetic pathways repress the activity of transposable elements (TEs), and so resetting epigenetic marks exposes the genome to mobilization of these elements, which may be particularly detrimental when producing gametes. In addition, meiosis may be specifically vulnerable to TE activity for several reasons. These include: (i) Deficient synapsis and repair due to the reshuffling of the meiotic machinery towards TE-induced DSBs; (ii) ectopic recombination among TEs, and (iii) interference with synapsis due to TE transcriptional activity. Alternative TE silencing mechanisms, such as those involving small RNAs, may have evolved to ensure proper TE control during epigenetic resetting. For example, these mechanisms involve piRNA and endo-siRNA in mammal male and female germlines, respectively, and transfer of siRNA from the central cell to the egg cell in plant female gametophytes. It is also possible that stringent synapsis checkpoints evolved, in part, to prevent the formation of defective gametes due to TE activity, along with other possible causes of meiotic errors.

Meiosis Asymmetry

Symmetrical meiosis results in four viable gametes, whereas asymmetrical meiosis results in a single gamete. Symmetrical meiosis is ancestral and is found in male meiosis in animals, seed plants, 'homosporous' species (e.g. mosses, many ferns) and isogamous eukaryotes. Asymmetrical meiosis, on the other hand, has evolved multiple times, and occurs in female meiosis in animals, seed plants and some ciliates. The selective scenarios underlying the evolution of meiotic asymmetry are unresolved. In some cases, such as in ciliates, there is no requirement for four meiotic products, as sex occurs by the cytoplasmic exchange of haploid micronuclei (conjugation). In other cases, asymmetrical meiosis in females results in a large oocyte full of resources, which may favour the production of a single cell rather than four. However, females could also achieve this symmetrically by undergoing fewer meioses. Therefore, is it possible that asymmetrical meiosis allows better control of resource allocation to oocytes, as symmetrical meiosis may not ensure an even distribution of resources across four meiocytes; one difficulty here is that it is not clear why female control of resource allocation would be more efficient among meiocytes derived from the same or different meiosis. A solution may be that meiocytes must compete for resources during meiosis, so that a symmetrical female meiosis is vulnerable to SGEs that bias resource allocation in their favour, possibly by killing other products of meiosis. Asymmetrical meiosis may therefore have evolved to suppress such costly competition within tetrads, it also opens the possibility of new conflicts.

Fairness of Meiosis

A striking feature of meiosis is its apparent fairness: under Mendel's first Law of inheritance, each allele has a 50% chance of ending up in any given gamete. However, there are many SGEs that increase their chances above 50% by subverting the mechanism of meiosis. These SGEs fall into two classes. The first class is killer SGEs, which kill cells that have not inherited the element. In principle, such killers could operate during meiosis (the hypothetical 'sister killers'), but the numerous killer SGEs that have been identified so far operate post-meiotically, e.g. by killing sibling sperm. The second class consists of meiotic drivers that exploit the asymmetry of female meiosis. These elements achieve transmission in excess of 50% by preferentially moving into the meiotic products that will eventually become the eggs or megaspores. There is a similarity between this kind of meiotic drive, where alleles preferentially go where resources are (i.e. the egg), and SGEs expressed later and biasing resource allocation in their favour. Parents make decisions of allocations to offspring before the 'meiotic veil of ignorance', whereas offspring compete for resources 'from behind the veil'. These genetic conflicts (between parent and offspring and between paternally and maternally derived alleles) are likely at the origin of parental imprints that differentially occur at male and female meiosis on some genes controlling embryo growth.

SGEs that undermine the fairness of meiosis provide explanations for otherwise puzzling observations. Perhaps most strikingly, centromere DNA regions often evolve rapidly, in contrast with what one would expect given their important and conserved function in meiosis. Henikoff *et al.*, therefore, proposed that expansion of repeat sequences in centromeric DNA produces a 'stronger' centromere, with increased kinetochore binding, which exhibits drive towards the future egg during meiosis I and, consequently, spreads in the population. Some of the best support for this hypothesis comes from a female meiotic driver in the monkeyflower *Mimulus guttatus*. Although conclusive evidence for a direct centromere function of this element is lacking, it is physically associated with large centromere-specific satellite DNA arrays. Female meiotic drive may also explain rapid karyotype evolution and the distribution of meta- versus acrocentric chromosomes because Robertsonian fusion chromosomes (fusions of two acrocentric chromosomes into one metacentric) can behave like meiotic drivers and segregate preferentially into the future egg during meiosis I.

Other features of meiosis may be adaptations to suppress killer or meiotic drive SGEs. Such adaptations are expected, because these elements are generally costly for the rest of the genome (e.g.). Defence against killer elements can be achieved by limiting gene expression. Accordingly, meiotic sex chromosome inactivation (MSCI, starting at pachytene of prophase I,) has been proposed to have evolved to control sex chromosome meiotic drive elements, and more generally this same principle may explain limited gene expression during meiosis and in its haploid products, as well as sharing of RNA and proteins among these cells. There is also evidence for rapid evolution and positive selection in the DNA-binding regions of centromere-associated proteins, which accords with the expectation of selection for countermeasures to limit preferential segregation of centromere drive elements towards the egg. The evolution of holokinetic chromosomes may be an extreme form of defence against centromere drive.

Meiosis and Recombination

A ubiquitous feature of meiosis is the exchange of genetic material between homologous chromosomes. The maintenance of recombination is even more debated.

The Number of Crossovers Per Chromosome: Constrained or Not?

In many species, the number of COs per bivalent appears to follow highly constrained patterns, showing little variation compared to the variation of chromosome sizes, themselves spanning several orders of magnitude. Within species, the correlation between genetic map length (in cM, with 50 cM being equivalent to 1 CO per bivalent) and physical length (in megabases, Mb) per chromosome is very strong ($R^2 > 0.95$), and often has an intercept of approximately 50 cM, consistent with occurrence of one obligate CO per bivalent. There is direct evidence indicating that bivalents lacking a CO have an increased probability of non-disjunction, resulting in unviable or unfit aneuploid offspring. Indeed, COs establish physical connections between homologues, promoting accurate disjunction by providing the tension needed for the bipolar spindle to establish. Therefore, this constraint has likely led to the evolution of regulation of CO numbers per bivalent across the eukaryotes. However, the reasons underlying the evolutionary persistence of this constraint are not well understood. In several species (e.g. *Arabidopsis*,), the intercept is less than 50 cM, but the smallest chromosome is at least 50 cM, thus still consistent with one obligate CO. More decisively, many species are achiasmate (i.e. have an absence of recombination) in one sex, with alternative mechanisms to ensure proper disjunction of achiasmate bivalents. This indicates that COs are not always obligatory and are maintained for reasons other than ensuring proper disjunction.

In addition to the obligate CO, additional CO events can occur within bivalents. The strong cM−Mb relationship within species indicates that the number of surplus COs correlates strongly with physical chromosome size. However, the rate at which surplus COs are added per Mb (i.e. the slope of the correlation) varies strongly between species. This may be partly explained by selection for different CO rates in different species. The strong correlations observed within most species may be explained by variation in *trans*-acting factors, such as the locus *RNF212* and its protein, which affects the propensity for DSBs to form surplus COs ; indeed, the identification of loci affecting variation in CO rates indicates the potential for rapid evolution of CO rates within and between species.

A further constraint on bivalent disjunction may exist: the separation of different bivalents on the meiotic spindle may need to be collectively synchronized to avoid aneuploidy. If the number of COs correlates with the amount of tension exerted on the homologues, then a tight control of excess COs may minimize disjunction asynchrony. This hypothesis may explain the observation that some disjunction problems in humans occur in a global manner without involving effects driven by specific chromosomes. Generally, high CO numbers are, on the other hand, not necessarily problematic with respect to proper disjunction.

Crossover Interference

A CO in one position may strongly reduce the likelihood of another CO occurring in the vicinity and on the same bivalent. This 'crossover interference' is widespread, but its function and mechanistic basis remains largely unknown. In many species, two classes of COs have been identified: Class I COs, which are sensitive to interference; and Class II COs, which are not. Class I COs are thought to play a major role in ensuring obligate COs, and so interference may limit the frequency or variance of COs, which may be important in ensuring proper disjunction. For instance, as with autopolyploids, increased interference may limit the number of COs to just one per chromosome, preventing aberrant multivalent segregation. A variant of this idea is that interference is a mechanism to avoid COs occurring in close proximity, which might reduce cohesion between homologues or slip and cancel each other out when they involve two or four non-interlocking chromatids, resulting in no CO occurring ; however, these mechanisms do not explain long-distance interference. A further suggestion is that CO interference may be adaptive by breaking up genetic associations. First, adjacent COs may be avoided because they cancel their effects on genetic associations. Second, it has been speculated that CO interference may reduce the chances of breaking up co-adapted gene complexes (supergenes). Some support for the idea that CO interference is not a purely mechanistic constraint comes from the fact that some species lack interference and, more importantly, that there is some evidence suggesting that interference levels evolve in long-term evolution experiments in *Drosophila*.

Differences in Recombination Rates between the Sexes

In many species, CO rates and localization differ between male and female meioses, and these differences can vary in degree and direction even between closely related species. The most extreme case is achiasmy, an absence of recombination in one sex, nearly always the heterogametic sex. This may have evolved either as a side effect of selection to suppress recombination between the sex chromosomes, or as a way to promote tight linkage without suppressing recombination on the X or Z chromosomes. More intriguing are the quantitative differences between males and females, known as heterochiasmy, which are found in many taxa, but whose mechanistic and evolutionary drivers are not yet fully understood. A number of explanations have been proposed, relating to mechanistic factors such as differences in chromatin structure, sexual dimorphism in the action of loci associated with CO rate (e.g. *RNF212*,), and evolutionarily widespread processes such as sperm competition, sexual dimorphism and dispersal. Some models point to a role of sex differences in selection during the haploid phase. While a viable explanation in plants, there is little empirical support for this in animals, where meiosis in females is only completed after fertilization (i.e. there is no true haploid phase), and where only few genes are expressed in sperm. However, meiotic drive systems are often entirely distinct between males and females and may be a primary cause of haploid selection. These systems often require genetic associations between two loci (a distorter and responder, or a distorter and a centromere in males and females, respectively). These driving elements might thus be very important in shaping heterochiasmy patterns. Indeed, COs in female meiosis are located closer to centromeres, which would be consistent with the view that this localization evolved to limit centromeric drive. Similarly, meiotic drive in favour of recombinant chromatids has been detected in human female meiosis, which may limit centromere drive.

Localization of Crossovers and Recombination Hotspots

The localization of recombination events differs between species. In many species, recombination occurs in localized regions known as 'recombination hotspots' of approximately 1–2 kb in length, although some species (e.g. *Caenorhabditis elegans* and *Drosophila*) lack well-defined hotspots. There are at least two types of hotspots. The first type, probably ancestral, is found in fungi, plants, birds and some mammals; these hotspots are temporally stable (up to millions of years) and concentrated near promoter regions and transcription start sites. The second type is likely derived and is found in other mammals, including mice and humans, where the positioning of hotspots is determined by the zinc-finger protein PRDM9. This system differs in two respects from the former: first, it appears to direct DSBs away from regulatory regions, and second, mutations in the DNA-binding zinc-finger array change the sequence motif targeted by the protein, leading to rapid evolution of hotspot positions over short time scales. This system is not present in all mammals: in dogs, hotspots target promoter regions, and the knock-out of Prdm9 in mouse makes recombination target promoter regions instead, underlining its derived nature.

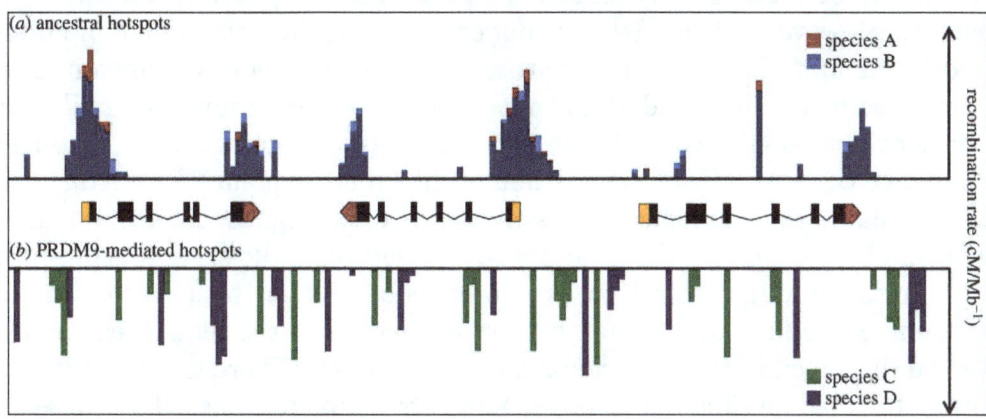

Hypothetical genome sequence containing three genes showing the distribution of ancient recombination hotspots in most model species (a) Compared with derived PRDM9-mediated recombination hotspots (b) Studies in fungi, plants, birds and dogs indicate that ancestral hotspots are stable over long evolutionary timescales (up to millions of years) and concentrate at promoter regions and transcription start sites (and at stop sites in some species). These start and stop sites for each gene are indicated in yellow and red blocks, with their introns and exons represented by lines and black blocks, respectively. PRDM9-mediated hotspots are found in some mammals, including humans and mice, and are directed away from promotor regions. The DNA-binding zinc-finger in the PRDM9 protein targets specific sequence motifs; mutations in the zinc-finger array change the targeted motif, leading to rapid evolution of hotspot positions and an absence of hotspot conservation over short evolutionary timescales (at the population and species level).

The evolutionary significance of both kinds of hotspots remains unclear. For the first type, the positions of hotspots may be caused by chromatin accessibility in transcribed regions or have evolved to favour recombination in gene rich regions (where it might be worth reducing genetic association). However, this does not clearly account for their precise location in regulatory regions. Another possibility might be that the co-occurrence of both COs and gene-conversion events (i.e. where resolution of DSBs without CO is achieved by exchanging small segments of DNA) specifically in regulatory regions could repress enhancer runaway, a mechanism that can lead to suboptimal expression levels. The evolutionary significance of the second kind of hotspot is similarly elusive. These hotspots are self-destructing because the target sequence motifs are eroded by biased gene conversion (BGC) during DSB repair.

This leads to a 'hotspot paradox': How can hotspots and recombination be maintained in the long term in the face of BGC? A possible solution is that *trans*-acting factors like PRDM9 may mutate sufficiently fast to constantly 'chase' new and frequent targets (hotspots), switching to new ones when these targets become rare due to BGC. This 'Red Queen' model does not require strong stabilizing selection on the number of COs, and closely mimics the pattern of hotspot turnover observed in some cases. However, this model does not explain how the second kind of hotspots evolved in the first place, as when it arose proper segregation was presumably already ensured by the first kind of hotspots (which, as seen in mice, are still active). Also, it does not explain why PRDM9 action is self-destructing: there is no necessity to induce DSBs exactly at the position of the target sequence for a *trans*-acting factor. In fact, there is no logical necessity to rely on a target sequence to maintain one CO per chromosome, as fixed chromosomal features could serve this purpose. It is worth noting here that recruiting promoter sequences for this purpose (as found for hotspots of the first kind) would be very efficient, as these sequences are highly stable and dispersed in the genome on all chromosomes. There is also no evidence so far that targeted binding motifs of PRDM9 correspond to some SGEs whose elimination would be beneficial. Overall, while spectacular progress has been made recently in elucidating hotspot mechanisms in detail (and patterns in recombination landscapes), there are still major gaps in our understanding of their evolutionary significance.

Significance of Meiosis

The significance of meiosis lies in providing gametes – the origins of next-generation progeny – with diverse gene combinations by mixing paternal and maternal genes. To create and transfer diverse combinations of DNA (genetic material) through the reproduction process is the key to ensuring the diversity of organisms.

In the meiotic process, homologous chromosomes form pairs, which are distributed independently to separate gametes. Gametes can therefore have many combinations of homologous chromosomes (shuffling of chromosomes). In this case, meiosis produces 2^3 or 8 types of gametes. For example, humans have 23 pairs of homologous chromosomes, and therefore the number of possible combinations is 2^{23} (8.4×10^6). As mentioned earlier, genetic crossing-over occurs between paired homologous chromosomes, resulting in gene recombination. Because crossing-over occurs independently in each sister chromatid, all four resultant chromatids have different gene combinations. In this way, through shuffling of chromosomes and chromosomal recombination, new sets of chromosomes with a mix of paternal and maternal chromosomes are created, generating diversity in gametes. Even if there are large numbers of sperm in the semen, none will be genetically identical to another.

Intraspecies genetic diversity is increased during the meiotic process through the formation of many homologous-chromosome combinations and gene recombination by crossover. This diversity is believed to be advantageous in creating progeny that can expand its habitat to a variety of environments and adapt to rapidly changing circumstances.

There are approximately 25,000 human genes. They are divided into 23 pairs of chromosomes, and shuffled through the reproduction process. Each chromosome will have approximately 1000 genes, so the genotypes and associated phenotypes on the same chromosome will be transferred as a set into the same gametes. This is called linkage. In some cases, however, genetic recombination can cause this linkage to break, in other words, the set of genes on the same chromosome can be separated and end up in different gametes.

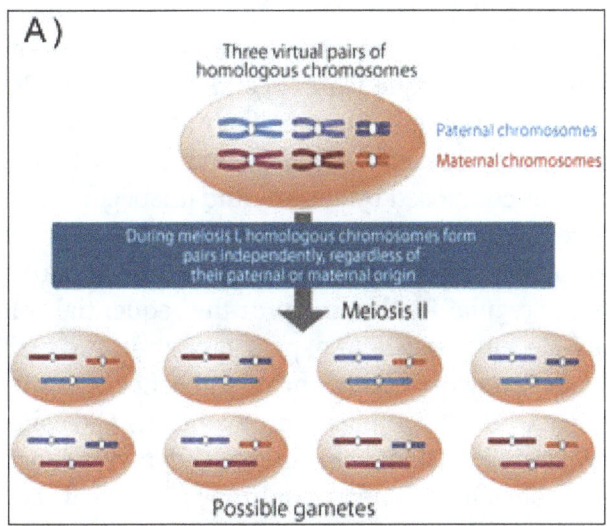

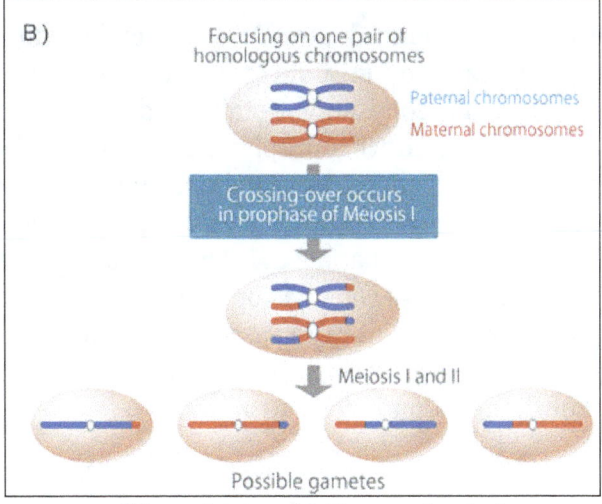

Models explaining significance of meiosis.

Cohesins and Cohesin-regulators in Meiosis

Cells have rigorous mechanisms controlling sister chromatid cohesion during cell division to ensure proper distribution of genetic material to daughter cells. Errors in these mechanisms often lead to aneuploidy, frequently implicated in cell death, tumour development and infertility. The main actor in this process is a four-protein complex called the cohesin complex. The role of cohesin complex in chromosome segregation is mediated by the formation of a ring-like structure, which entrapped replicated DNA. The dynamic of the cohesin ring is regulated by a still undetermined number of cohesin-interacting proteins. These cohesin-regulators were essentially identified and studied in relation with the cohesion function of cohesin complexes. In the last years, we have improved our understanding of the key players in the regulation of sister chromatid cohesion during cell division in mitosis and meiosis. During meiosis the formation and disassemble of synaptonemal complex (SC), the recombination between homologue chromosomes and the maintenance of sister chromatid cohesion until metaphase II - anaphase II transition require unique features and players participating in the control of chromosome segregation. While in mammalian mitosis there are essentially two cohesin complexes, which only differ in the STAG subunit (STAG1 or STAG2), in meiosis several cohesin complexes composed by specific and unspecific meiotic cohesins coexist and probably they develop different functions depending of their spatio-temporal chromosome localization. On the other hand, a correct control of chromosome cohesion in meiosis involves specific regulators that monitor the sequential cohesin release during both meiotic divisions. Recently excellent papers have appeared in the literature looking in depth on the molecular control of sister chromatid cohesion and cohesins in cell division. In this chapter, we review the implication of cohesins and cohesin-interacting proteins in meiosis-specific processes and chromosome dynamic. Furthermore due to the increasing relevance of cohesins in human syndromes, we briefly point how problems in their tasks during mammalian mitotic and meiotic cycle drive to pathological situations.

The pair of sister chromatids produced after DNA replication must be maintained together throughout the G2 phase and until its segregation to ensure a correct cell division. Thus, the sister chromatid cohesion is established during S phase and although in mitosis the cohesion is released during prophase and prometaphase in arms, the sister chromatid are joined at centromeres until the onset of anaphase. In meiosis, there are two consecutive chromosome divisions, segregating homologous chromosomes in anaphase I and sister chromatids in anaphase II. Thus the cohesion regulation presents specific characteristics of this kind of cell division, and whereas the arm cohesion is lost at anaphase I, the centromere cohesion is released at anaphase II. Since both cohesion releases follow the same pathway in meiosis, the meiotic centromere cohesin complexes must be protected until the second meiotic division.

At any case, mitosis and meiosis, the multi-protein complex responsible of sister chromatid cohesion is called cohesin complex and it was first characterized in Saccharomyces cerevisiae and Xenopus laevis. The cohesin complex is composed by four subunits: Two structural maintenance of chromosomes family proteins (SMC1 and SMC3), one α-kleisin subunit (SCC1/RAD21), and a HEAT-repeat domain protein (SCC3/SA/STAG). The most of components of the cohesin complex are conserved from yeast to humans but there are specific subunits depending on the species or the cell division type. The most conserved cohesins are the SMC proteins, which form a V-shaped heterodimer, representing the core of the cohesin complex. Higher eukaryotes have two mitotic SA/STAG family members, SA1/STAG1 and SA2/STAG2, which do not coexist and are present in different cohesin complexes. In germ cells have been characterized distinct meiosis-specific subunits of cohesin complex in different organisms.

In mammals, a meiotic paralogue of SMC1 has been described, the SMC1 β, thus the subunit presents in mitosis and meiosis is called SMC1 α. In yeast and mammals REC8 is the meiotic paralogue of SCC1/RAD21 subunit and in mice a new α-kleisin has been identified recently, the RAD21L, a paralogue of RAD21. RAD21L interacts with STAG3 (meiosis-specific SCC3 subunit) and with the three described SMC cohesin subunits, SMC1 α, SMC1 β and SMC3. STAG3 is the meiosis-specific paralogue for STAG1/2 in mammals. Whereas these cohesin subunits, RAD21L, REC8, STAG3 and SMC1 β, are meiosisspecific and they are present specifically in spermatocytes and oocytes, different cytological and molecular analysis show the participation of the SMC1$^\lrcorner$, RAD21 and STAG2 in mammalian mitosis and meiosis. Characterization of distinct meiotic cohesin complexes containing REC8, RAD21 or RAD21L as $^\lrcorner$-kleisin subunits, which have distinct localization patterns and dynamics, as well as the simultaneous presence of SMC1 α and β-containing complexes and the presence of STAG2 and STAG3 in mammalian meiosis suggest a large variety of putative cohesin complexes formed by combinations of cohesin subunits.

The most accepted model of the cohesin complex organization describes a heterodimer of SMC proteins jointed by their hinge domains. The ATPase heads of SMC1 and SMC3 are connected by the α- kleisin, forming a tripartite ring and finally the SCC3 subunit interacts with SCC1 via SCC1's C-terminus. Two major models, which are not mutually exclusive, have been proposed for the cohesin complex function in sister chromatid cohesion and its interaction with the DNA molecules. The first one is based on the electronic microscopy results and structural characteristics of SMCs and described a cohesin complex forming a ring-like structure, which mediates cohesion by embracing chromatin fibers of both sister chromatids. The second model was named handcuff model, it involves the participation of two cohesin rings formed by SMC1/SMC3/SCC1 subunits which interacts in an SCC3-dependent manner. In this model each sister chromatid is encircled by a tripartite cohesin ring.

The activity of the cohesin complex is closely related to the action of three cohesin cofactors: PDS5, WAPL and Sororin. PDS5 is associated with cohesins but in a less tightly bound manner than the cohesin complex proteins. In vertebrates there are two PDS5 homologues, PDS5A and PDS5B. The role of PDS5 is related to the maintenance of cohesion and the modulation of the interaction of cohesin complex with the chromatin. WAPL is involved in heterochromatin organization in Drosophila melanogaster and in human cells regulates the resolution of sister chromatid cohesion during prophase.

A. The cohesin complex is formed by four proteins: SMC1, SMC3, SCC1 and SCC3 (green, red, orange and blue piece respectively). The SMC1, SMC3 and SCC1 subunits form a ringlike structure. The SCC3 protein interacts with SCC1 to complete the cohesin complex. The cofactors PDS5 (yellow piece) and WAPL (purple piece) interact to form a protein complex, which is associated to the cohesin complex by SCC3 interaction. B. Ring-embraced model of cohesion. A single cohesin complex hugs the two sister chromatids. C. Handcuff model of cohesion. Two cohesin rings interacts by a single SCC3 molecule. Each cohesin ring encircles a single chromatid.

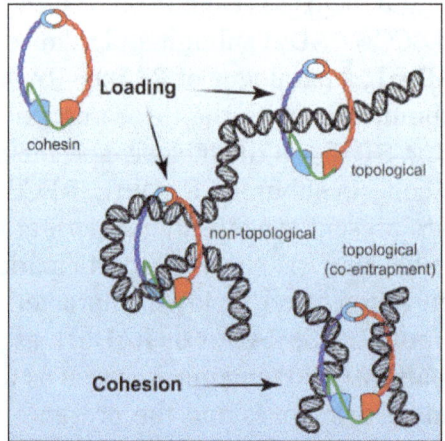

Cohesin complex organization and cohesion models.

Regarding to the Sororin, it has been described in culture cells that this protein have a role in centromere cohesion, probably it is implicated in the protection of centromeric cohesin complexes. Moreover Sororin seems to have a role in the cohesion establishment during replication. The implication of all these cofactors in meiosis is a wide field of study nowadays. The importance of cohesins, cohesin complex and cohesion cofactors not only consists in keeping jointed both sister chromatid and then ensure the bipolar attachment and the chromosomal segregation in mitosis and meiosis, but the cohesins acquire a pivotal role in many different aspects of meiosis and chromosome dynamics such as the modulation of gene expression, the double strand breaks (DSBs) repair, axial elements (AEs) formation or pairing and synapsis of homologous chromosomes. In this chapter some of those different roles of cohesins and cohesion cofactors are reviewed, giving special attention to the specificity of some cohesin subunits to specific functions.

Loading, Establishment, Maintenance and Release of Cohesin Complexes

Chromosome segregation control is managed by an intricate network of events, which assures that each daughter cell receives the right chromosome number, equal amount of chromosomes in mitosis and a haploid recombined chromosomal pack in meiosis. On one hand, fails in mitotic chromosome segregation compromises the viability of the organism, on the other hand fails in meiotic chromosome segregation jeopardizes the fertility of individuals. The importance of cohesion must be taken into account from the loading of cohesins to the establishment of cohesion, maintenance of cohesive function and finally the removal of cohesin complexes. In this context, the association/dissociation of sister chromatids is a wellcontrolled process that involves the role of different cohesin-interacting proteins. The loading of cohesin complexes to chromosomes occur near the G1 to S phase previous to cell division, thus it would be easy to think that the regulators of cohesin loading would be similar in mitosis and meiosis. In fact, most of them have been first described in yeast mitosis and afterwards a similar meiotic function was identified in some cases. Although we do not should forget that the transition from mitosis to meiosis could be trigger from the premeiotic S phase, and taking into account the presence of meiotic cohesin complexes and the special regulation of meiotic cohesion release, crucial differences might exist between mitosis and meiosis. It was described in S. cerevisiae that the loading of cohesin complexes depends on the SCC2/SCC4 adherin complex.

This loading is not sufficient for the cohesion function in chromosome segregation and the Eco1/Ctf7p acetyltransferase is required for the establishment of cohesion in mitotic yeast chromosomes. In humans, two Eco1 orthologues have been identified: ESCO1 and ESCO2. Both proteins exhibit not redundant acetyltransferase activity and the depletion of any of them by siRNA cause defects in sister chromatid cohesion and chromosome segregation. However, in S-phase the role of both proteins seems to be directly related to the cohesion establishment and do not with the cohesin loading. The substrate of Eco1/ESCO1 is SMC3 in both human and yeast cells, being required the acetylation of SMC3 for sister chromatid cohesion. The acetylation of SMC3 could regulate its interaction with SCC1, promoting the turn from a chromosome-bound cohesin complex into a cohesive complex, suggesting that the cohesin complexes with acetylated SMC3 may be stably bound to chromosome. The activity of Eco1 could be facilitated due to its interaction with PCNA (proliferating cell nuclear antigen). The acetylation of SMC3 is maintained until metaphase - anaphase transition and its deacetylation in yeast mitosis depends on Hos1, but this deacetylase cannot act until the SCC1 cohesin subunit is cleaved by a protease called Separase, recycling then the tripartite open ring of cohesin. In mammalian mitosis the deacetylase responsible for SMC3 deacetylation has not been identified, moreover the cycle of SMC3 acetylation/deacetylation in meiosis is still poorly known.

The release of cohesion in mitotic prophase and anaphase follows different pathways of resolution. In mitosis, the arm cohesion is released during prophase in a Separase-independent manner, it depends on the phosphorylation of SCC3/SA2/STAG2 subunit by Aurora B and Polo-like kinases, but the correct chromosome segregation depends on the centromeric cohesion, which is maintained until anaphase onset and then released by the Separase activity. The release of cohesin complexes from centromeric chromatin at mitotic metaphase - anaphase transition is mediated by the Separase, a specific protease that cleaves the SCC1 subunit of the cohesin complex, destabilizing the association of cohesins to chromatin. Before anaphase, Separase remains inactivated by binding to its specific inhibitor Securin. Activation of the anaphase promoting complex/cyclosome (APC/C) leads to ubiquitination of Securin, allowing cleavage of SCC1/RAD21 by Separase and triggering the onset of anaphase. In meiosis there are two consecutive chromosome segregations which are triggered by a Separase-dependent mechanism of lost of cohesion. During first meiotic division, the cohesin complexes at chromosome arms are removed during the metaphase I - anaphase I transition, allowing segregation of recombined homologues to opposite poles. In the second meiotic division, the cohesion is released from centromeres and sister chromatids are segregated to opposites poles to generate haploid gametes. Then, this type of cell division presents unique characteristics in sister chromatid cohesion removal and centromere cohesion protection. This is necessary to prevent premature separation of sister chromatids in order to avoid aneuploidy in the resulting gametes. Related to the centromere protection a protein family was indentified in fission yeast, the Shugoshins: Sgo1 and Sgo2. In these organisms Sgo1 seems to be essential in meiosis and Sgo2 is mostly implicated in mitotic division. The activity of Sgo1 and Sgo2 is related to the recruitment of a serine/threonine protein phosphatase 2A (PP2A). In humans these proteins are called SGOL1 and SGOL2 and in mitosis they also collaborate with PP2A. In mice it has been described the localization of SGOL2 in male meiosis and it corresponds with a protection function of centromeric cohesion during first meiotic division. An interesting model is the SGOL2- deficient mouse, where a precocious dissociation of the meiosis-specific REC8 cohesin complexes from anaphase I centromeres was observed, demonstrating the specific implication of SGOL2 in centromere protection during meiosis.

The cohesin-regulators PDS5 and WAPL are also involved in the opening/closing of cohesin ring by interactions with different cohesin subunits. These cofactors are not required for cohesin association to chromosomes but they are necessary for cohesin complex dynamics. The action of PDS5 is related to its interaction with the SA1 and SA-containing complexes in somatic cells. In vertebrates two PDS5 proteins have been characterized: PDS5A and PDS5B. Both are large HEAT-repeat proteins that bind to chromatin in a cohesin-dependent manner in human cells and Xenopus egg extracts. RNAi depletion of PDS5A and PDS5B show that both are needed for maintaining cohesion, altering preferentially centromeric cohesion in Xenopus egg extracts. Mice lacking PDS5B function die shortly after birth and exhibit multiple developmental anomalies that resemble those found in humans with Cornelia de Lange syndrome (CdLS), indicating a relevant function for PDS5B beyond chromosome segregation, but there are no discernible defects in sister chromatid cohesion. Despite of these contradictory results, the function of PDS5 has been related with the maintenance of sister chromatid cohesion during G2. Although, whether PDS5 also has a function in sister chromatid resolution remains to be determined. Another interesting cohesin cofactor is the product of the previously identified Drosophila wings apart-like (Wapl) gene, involved in heterochromatin organization. Human WAPL regulates the resolution of sister chromatid cohesion and promotes cohesin complex removal by direct interaction with the RAD21 and SA/STAG cohesin subunits. Thus, WAPL seems to destabilize cohesins. It has been proposed that PDS5 and WAPL form a protein complex, which in association with SCC3 cohesin subunit antagonizes the establishment of cohesion, calling the WAPL-PDS5 anti-establishment complex.

Skibbens proposed a model of cohesion establishment where the recruitment of PDS5-WAPL complex onto SCC3 subunit prevents the binding of cohesin complexes by destabilizing cohesin-cohesin association. During S phase, Eco1/ESCO1 acetylates SMC3, which inhibits the WAPL-PDS5 activity temporally. After S phase, PDS5 would stabilize the cohesin complexes activity.

Sororin has been implicated in centromere cohesion. This protein was firstly identified in a screen for substrates of the APC in vertebrates and no homologues have been described in other organisms. Different results in somatic cells suggested that Sororin interacts with the cohesin complex and it is essential for the maintenance of sister chromatid cohesion. Sororin is ubiquitinized and degraded after sister chromatid cohesion is dissolved. Studies on Sororin-depleted and Shugoshin-depleted cells indicate that both proteins might act in concert in the protection of centromeric cohesion. Sororin is also needed for maintaining stable chromatin-bound cohesin and DSBs repair in G2. The Sororin recruitment depends on Eco2/ESCO2, both are subtrates of APC and its activity is related to the DNA replication.

In agreement with these findings, Nishiyama reported that DNA replication and cohesin acetylation promote binding of Sororin to cohesin complex and that Sororin displaces WAPL from its binding partner PDS5, thus it would contribute to maintenance of a stable binding of cohesin to chromatin. Despite of the relevant roles in chromosome cohesion control suggested by all these results, there are few data regarding to the putative functions of cohesion cofactors such as SCC2/SCC4, Eco1/ESCO1, Eco2/ESCO2, WAPL, PDS5 or Sororin in chromosome segregation in meiosis.

Roles of the Cohesin-regulators in Meiosis

Updates there are no many evidences regarding to the specific implication of all the cohesion cofactors in meiosis. It is predictable that all of them might act in meiosis in a similar way to that described in mitotic cells or perhaps they acquire specific roles due to the particular regulation of meiotic division. Update the unique indication that ESCO2 have a role in meiosis was the identification of Esco2 gene as a candidate to be a potential regulator of the transition from mitosis to meiosis in mammals, identifying by means microarray database in both testis and ovary of mouse. The data are supported not only by a pattern of mRNA expression but also by protein immunolocalization, showing on one hand that Esco2 is expressed in testis and ovary, specifically in the embryonic gonad, and on the other hand that ESCO2 localized to the nucleus of spermatocytes at early prophase I. This last probe was performed by immunofluorescence techniques over testis sections at different ages. At 10dpp (days post-partum) male mice, ESCO2 was presented at preleptotene and leptotene diffusely and from 15dpp this protein is located at pachytene in a discrete domain within the nuclei. Our laboratory is actually studying the ESCO2 implications in meiosis and using the same antibody (Bethyl Laboratories, A301-689A) we have identified that the discrete domain observed by Hogarth et al., is in fact the XY body figures. We have analyzed the male mice meiosis over squashes of seminiferous tubules and spreads of meiocytes and performed the doubleimmunolocation of ESCO2 and SYCP3. Our results indicate that the ESCO2 acetyltranferase is present over the chromatin of XY body from zygotene to late stages of prophase I, detecting a more intense signal at pachytene. We distinguish the XY body in squashed spermatocytes because of it usually is located at nuclear periphery and its AEs are partially unsynapsed due to this pair of chromosomes share only a homologue region called pseudoatosomal region, which the SC is exclusively formed in. Also the DAPI chromatin staining led us to identify the XY body undoubtedly in squashes. At pachytene nucleus the synapsis between autosomes is completed, which can be detected by SYCP3 immunolabeling.

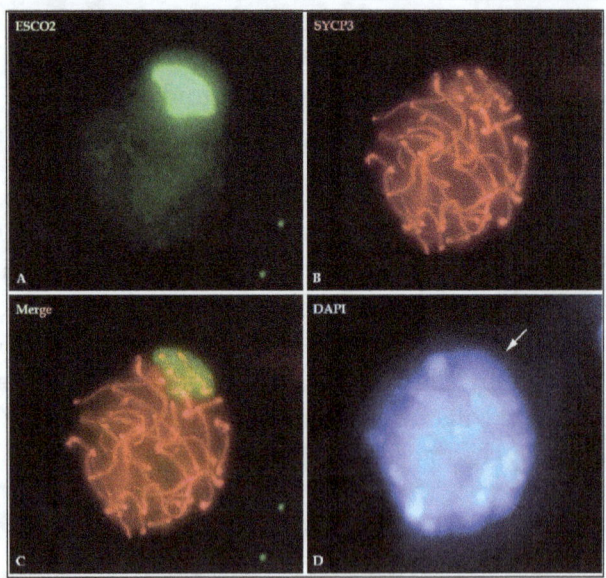

Immunolabeling of ESCO2 (green) and SYCP3 (red) at pachytene after squasing of spermatocytes.

A. ESCO2 immunolabeling (rabbit anti-ESCO2 Bethyl Laboratories, A301-689A; 1:50 dilution) mark the XY body at pachytene. FITC-conjugated anti-rabbit IgG secondary antibody (Jackson Laboratories) was used at 1:150 dilution. B. SYCP3 (mouse monoclonal antibody anti-SYCP3. Santa Cruz Biotechnology, SCP-3 (D-1): sc-74569; 1:100 dilution) mark the LEs of synapsed autosomes and the AE/LEs of the XY body. DyLight594-conjugated anti-mouse IgG secondary antibody (Jackson Laboratories) was used at 1:150 dilution. C. The merge image of ESCO2 and SYCP3 signals is shown. D. The chromatin was counterstaining with DAPI in blue (4',6-diamidino-2-phenylindole, SIGMA). The XY body is indicated (white arrow). After mounting the preparations with Vectashield (Vector Laboratories), the meiocytes were visualized using a Leica AFX6000LX multidimensional microscopy. The images were captured with LAS_AF software and analyzed and processed with public domain ImageJ and Adobe Photoshop CS3 softwares. All images are the result of superimposition of all the focal planes occupying the total volume of a pachytene mouse spermatocyte after squashing of seminiferous tubules.

We have not detected any ESCO2 labelling over the autosomas at any stage of meiosis. Since the X and Y chromosomes in all eutherian mammals preserve only a small region of homology, the genetic and morphological differentiation of sex chromosomes mark the XY behaviour during meiosis, which cannot be comparable with autosomes. Thus, structural modifications in the unsynapsed AEs of XY pair have been identified. Moreover, modifications in the distribution of different cohesin subunits have been also observed, as the preferential location of REC8 in the synapsed region of the XY body at pachytene. However, our results show that the ESCO2 labelling embrace all the chromatin of sex chromosomes, similar signals have been detected after immunostaining of (γ-H2AX) and surprisingly the cohesin subunit RAD21L has been also observed in the XY chromatin at pachytene and diplotene.

The presence of RAD21L over the XY chromatin has been explained as part of the sexual dimorphism observed regarding to the detection of RAD21L in mice and the authors pointed to a specific role of this meiotic cohesin subunit in the pairing and development of the sex body, although deep studies should be performed to understand the role of a cohesin subunit in all sex chromatin. (γ-H2AX) is the histone variant derived from the phosphorylation of H2AX at serine 139 as consequence of DSBs, however the presence of this chromatin modification in XY pair has been related to the transcriptional repression associated with the sex body.

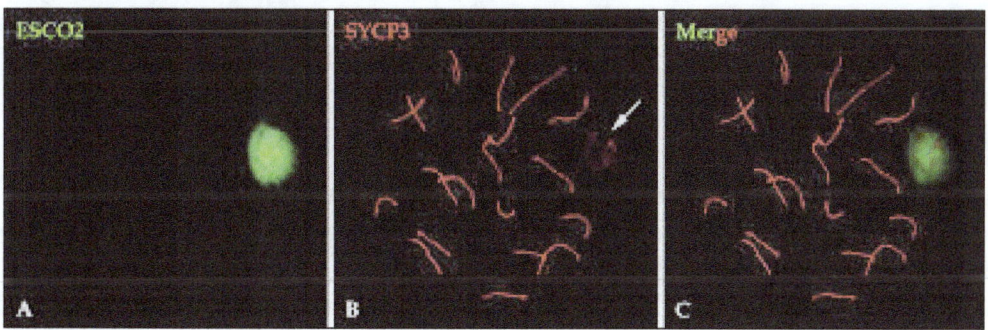

Immunolabeling of ESCO2 (green) and SYCP3 (red) at pachytene after spreading of spermatocytes.

A. The immunolabeling of ESCO2 (rabbit anti-ESCO2 Bethyl Laboratories, A301-689A; 1:50 dilution) mark the XY body at pachytene. FITC-conjugated anti-rabbit IgG secondary antibody (Jackson Laboratories) was used at 1:150 dilution. B. The LEs of synpased autosomes and the AE/LEs of the sex chromosomes (white arrow) are detected with an antiSCYP3 antibody (mouse monoclonal antibody anti-SYCP3. Santa Cruz Biotechnology, SCP-3 (D-1): sc-74569; 1:100 dilution). DyLight594-conjugated anti-mouse IgG secondary antibody (Jackson Laboratories) was used at 1:150 dilution. C. Merge image of ESCO2 and SYCP3 signals. After mounting the preparations with Vectashield (Vector Laboratories), the meiocytes were visualized using a Leica AFX-6000LX multidimensional microscopy. The images were captured with LAS_AF software and analyzed and processed with public domain ImageJ and Adobe Photoshop CS3 softwares.

During male meiosis, X and Y chromosomes are silenced in a process called Meiotic Sex Chromosome Inactivation (MSCI), it has been described that this chromosome inactivation is essential for male meiosis progression, thus the disruption of MSCI arrest the male meiosis at pachytene, being essential for male fertility. According to the formation of XY body and its transcriptionally silenced status during spermatogenesis different histone modifications have been described at pachytene such as H3 and H4 deacetylation or H3-K9 dymethylation. The specific function of ESCO2 in the sex chromatin of male mice is not known. However it has been described that this acetyltransferase could play a role in regulating transcription, specifically it has been suggested a transcriptional repression activity through the interaction of ESCO2 with chromatin modifying enzymes and with the CoREST complex, achieving the repression by means of histone modification. Moreover, the authors talk about an ESCO2- containing complex, which has histone methyltransferase activity in culture cells. These evidences could link the presence of ESCO2 in sex chromatin and the MSCI process. Another cytological example of the specific presence of a cohesion cofactor in meiosis is related to WAPL localization. WAPL was found on axial and lateral elements of chromosomes (AE/LEs) in some prophase I stages in mouse spermatocytes and oocytes (Zhang et al., 2008c) colocalizing with SYCP3, a constitutive protein of the SC; however, no more extensive study has been carried out on the role of WAPL in meiosis. It has been established that WAPL and PDS5 act as a protein complex during cohesion establishment and maintenance; but there is no data about the possible presence of PDS5 in mammalian prophase I.

However in budding yeast meiosis has been described that Pds5 is required for pairing of homologous chromosomes and this cofactor could have a role inhibiting the synapsis between sister chromatids. Depletion of Pds5 gives raise to fails in synapsis, hypercondense chromosomes and blocking meiocytes at pachytenelike stage. This result was preceded by the Pds5 fission yeast mutant that showed absence of SC formation. The action of Pds5 in pairing and synapsis could be related with the interaction between both proteins and the Pds5 capacity of modulating Rec8 activity in both fission and budding yeast. Although these studies point out to the importance of the structural organization of meiotic chromosomes and the implication of cohesin complexes and cohesion cofactor in such important process as synapsis of homologous chromosomes, further cytological and biochemical studies are needed to characterize the role of PDS5 and WAPL during meiosis. Are those cohesion cofactors acting in unknown manner or their activity is related to the cohesion modulation? This is an unresolved question yet. Finally a recent molecular-genetic approach have shown that the cohesin-loading factor Scc2 and the subunits Scc3 and Smc1 of the cohesin complex are required for activating the production of the meiosis-specific subunit Rec8 in the S. cerevisiae. This result suggests that Scc2 could play a dual role in gene regulation and sister-chromatid cohesion during meiotic differentiation. Although the mechanism is poorly understood yet, the role of cohesins and cohesion cofactors in gene regulation has been largely studied in the somatic line. However, meiosis-specific gene regulation is a future field of study where probably different groups are work in.

To Form Bivalents: Cohesion, Synapsis and Recombination

Two of the most important phenomena that characterize the meiosis are the SC formation and the reciprocal homologous recombination. Both are meiosis-specific and are implicated in the physical connexion of homologous chromosomes to form and maintain the bivalents. In this context, the cohesion is the third pivotal factor. Since the cohesion stabilizes the joint between sister chromatids from S phase until their segregation, it is easy to think that the cohesin complexes are needed to ensure that both sister chromatids of each homologous chromosome act in a single manner during prophase I, in a cellular context where the AEs formation, the synapsis and the meiotic recombination between homologous chromosomes is taking place. This would be the role of cohesin complexes in sister chromatid cohesion, its canonical function. Thus, during first meiotic division these three events: cohesion, synapsis and recombination, must be perfectly coordinated. However, the existence of meioticspecific cohesins and its specific localization during first meiotic division become evident that the meiotic cohesin complexes could be implicated in SC formation and recombination directly further than the canonical function of chromatid cohesion.

Importance of Individual Cohesin Subunits to Synapsis

The SC is a meiosis-specific proteinaceous structure formed by two lateral elements (LEs), derivate from AEs of chromosomes, and a central element (CE) connected by transverse filaments (TFs). It is known that this meiotic structure play important roles in the condensation and pairing of chromosomes and also has a close relationship to the programmed DSBs formation and resolution. The formation of SC between homologous chromosomes is highly conserved in evolution, it derivates in the synapsis of homologous chromosome during the meiotic prophase I. One of the first references of the close relationship between the SC and cohesion was observed in budding yeast. In these organisms the cohesin subunit Smc3p presents a continuous localization along chromosome cores similar to that described for SC proteins in rat. Klein et al., included a functional analysis, they studied the Smc3 and Rec8 mutants and observed that both were defective in cohesion, formation of AEs and SC assembly, demonstrating in yeast that the cohesins were essential not only for the meiotic sister chromatid cohesion but also for the synapsis. Moreover they postulated that the meiotic cohesin complexes were in fact part of AEs of chromosomes. However, two years later Pelttari et al., went farther and proposed that the cohesin axis was preformed along each chromosome and might act as the organizing framework for AE/LEs formation.

This proposal was based on the evidences that Sycp3-/- mice were able to form a co-hesin axis even in the absence of AEs. Contrary, the Rec8-/- and Smc1β-/- mice showed AEs partially assembled. In rat spermatocytes the interaction between SMC cohesin subunits and SYCP2 and SYCP3, the main components of AEs and LEs, was also stud-ied, concluding that the cohesin axes were essential in the SC formation and synapsis progression during early stages of prophase I. These first observations pointed to the contribution of cohesins not only to form the AEs but also as part of AEs/LEs and then to its function in synapsis. Since then we have known different examples which closely relate the SC proteins and the synapsis progression to cohesin localization and function in meiosis. In the table we summarize some of examples of the contribution of cohesin to SC formation and synapsis that we detail forward.

The evidences of the implications of cohesin complexes in SC assembly, pairing and AEs formation came from many different organisms and scientific approaches. In plants, a specific and intermittently localization of SMC3 in the AE/LEs has been observed in to-mato by means immunogold labeling and electron microscopy in zygotene microsporo-cytes, similar to that observed by light microscopy after the immunolabeling of SMC1, SMC3, SCC3 and REC8, although no all subunits presented the same pattern of accu-mulation and appearance during prophase I. In this vegetal species REC8, SMC1 and SMC3 are localized as foci from preleptotene which are arranged to AEs at leptotene but do not colocalize. However they exhibit colocalization in synapsed regions at zygo-tene and pachytene. SCC3 appears later than the other three cohesin subunits, observ-ing a substantial amount of SCC3 foci along AE/LEs from zygotene through pachytene, according with the SC assembly. There is no a functional analysis about the implication of this vegetal cohesin in SC formation and synapsis, but this sequential loading shows a possible change of cohesin complexes composition or organization at a meiotic stage which is key in the correct progression of synapsis.

This non-homogenous timing and spatial localization of different cohesin proteins has been reported in other organisms such as C. elegans or male grasshoppers. Males of grasshoppers Locusta migratoria and Eyprepocnemis plorans species represent anoth-er cytological example of the contribution of cohesins to synapsis. In grasshoppers the SMC3 subunit appeared as a dotted signal in preleptotene to form continuous lines in leptotene, meanwhile the non-SMC subunits RAD21 and SA1 were not localized until zygotene and only on synapsed regions. This means that in grasshoppers the incor-poration of non-SMC cohesins would be spatially and temporally concomitant with the progression of synapsis. In zygotene and pachytene SMC3 and RAD21 present a continuous and linear pattern, similar to what would be expected of AE/LEs proteins and both cohesin subunits present a complete colocalization until diplotene, coincid-ing with the stages of disassemble of SC. It is relevant that the single X chromosome of grasshopper males, which remains unsynapsed, presents neither RAD21 nor SA1 during prophase I. Further than the possible implication of RAD21 and SA1 in the complete assembly of SC or vice-versa, these observations open new questions: Would this synapsis-dependent sequential loading of cohesin during prophase I have an effect on the distribution of cohesins in metaphase I? Could we talk about a presence of cohesin depending on the synaptic history of chromosomes? At any case, these kind of studies pointed not only that cohesin subunits can be loaded onto meiotic chromosomes at different time or as part of different complexes, but also relate the loading of specific cohesin subunits with the timing of AEs formation and SC assembly. The idea of specific synapsis cohesin complexes is also supported by the evidence that SMC1 α is mainly located along synapsed regions in mice spermatocytes, whereas SMC3 subunit localizes along the synapsed and unsynased AEs, in front of SMC1 β which is mainly required for sister chromatid cohesion.

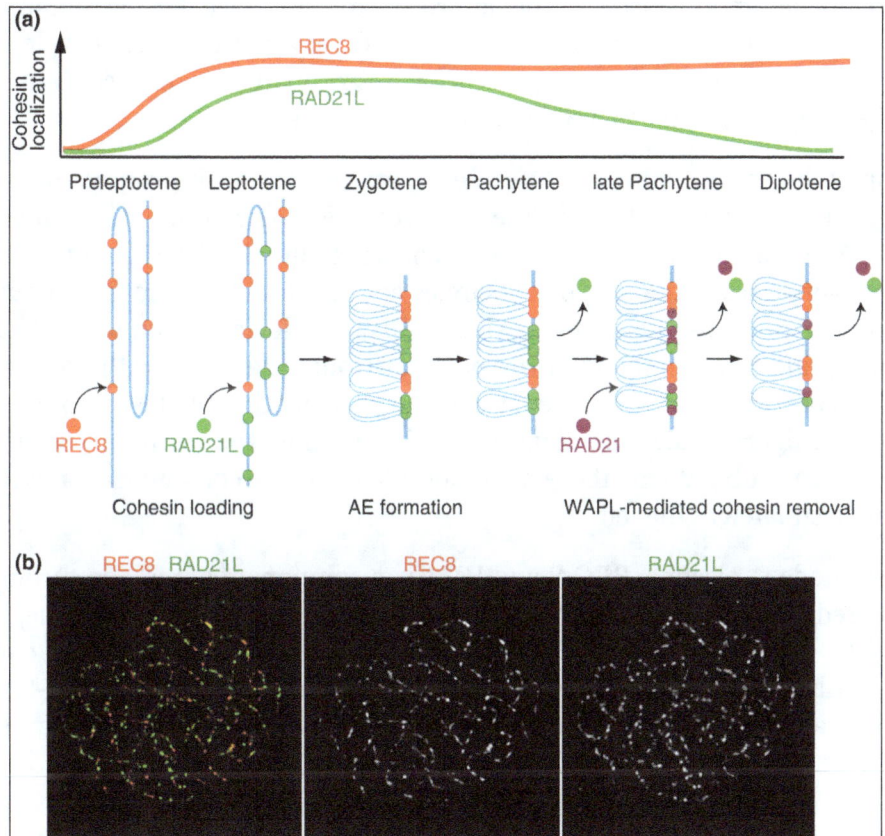

Temporal meiotic appearance of cohesin subunits from preleptotene
and its relationship with SC formation.

In addition, Sycp3-/- and Smc1β-/- single and double oocytes mice mutants show that both proteins contribute to meiotic chromosome axis organization. The authors propose a several layers axis association, which contributed to the formation of meiotic chromosome axes in an independent manner. The basic layer would be formed by the same proteins that form the mitotic chromosome axis, the scaffold proteins. SMC1 α would be part of the second layer, it contributes to axes length but is not essential. Finally SYCP3 would represent the third layer as basic component of AE/LEs. Thus a cohesin axis would be defined as a base axis before the AE/LEs in mammals, a similar organization has been proposed in grasshoppers. Smc1β -/- mice spermatocytes do not progress further than midpachytene and the cells present abnormalities in chromosome structure. The analysis of SC proteins showed that the AEs were shortened, but the synapsis occurred between the 19 couples of homologous chromosomes. Thus, despite of the close relationship of the cohesins and SC proteins and the evident interdependence between both groups of proteins, there are examples in mice where the absence of a cohesin subunit does not inhibit the complete formation of SC, albeit its structure is compromised. This could be explaining due to the presence of different cohesin complexes in mammal meiotic cells which could act in a redundancy manner. All these evidences suggest that the cohesins and specifically the SMC cohesin provide a basis for AE assembly. Based on the description of the new cohesin subunits and regarding to its role in SC formation and synapsis, diverse authors have develop models of incorporation of new cohesin complexes or replacement of specific cohesin subunits during the early prophase I that we summarize in figure.

At the top of the figure a temporal line of early stages of meiosis progression is shown. The diagram positioned under this line is represented the cohesin axis formation and the SC assembly taking into account the data observed in mice. The blue lines represent the chromosomic scaffold and the blue circles are the chromatin loops. The pink color represents the cohesin organization, which is usually detected as a dotted pattern at preleptotene and from leptotene form axis that would the base for the AEs proteins. The AE/LEs are represented in green, the FTs in red and the CE of SC in purple. At the bottom of the figure, under the diagram of SC assembly, the temporal localization of different cohesin subunits in different species is shown. The colour code for each cohesin subunit is situated at the left.

In the nematode C. elegans, REC-8 and HIM-3, a component of meiotic chromosomes axes required for synapsis, overlap during leptotene, zygotene and pachytene, suggesting that REC-8 is a component of AE/LEs. SMC-1 and SMC-3 loaded onto chromatin independently of REC-8 and SCC-3. On the other hand the depletion of Rec-8 by RNAi did not cause change in SMC-1 or SMC-3 localization but a mislocalization of SCC-3 was detected. The importance of REC-8 as an associated component of AEs in C. elegans derives from the evidence that this protein is presented along the axes of unsynapsed chromosomes and this meiosis-specific cohesin subunit seems to be necessary for the assembly of SC components such as HIM-3, SYP-1 and SYP-2. Moreover, the analysis of Rec-8 (RNAi) mutants showed that this protein is implicated in the chromosome pairing and SC assembly during leptotene/zygotene stage and the meiotic chromosomes are reorganized before diplotene into pachytene-like arrangement but they fail to achieve a normal lengthwise alignment. Probably REC8 is the subunit of cohesin more widely studied in meiosis, it is a meiotic-specific cohesin variant of the subunit SCC1/RAD21. REC8 has orthologs in different organisms, it is present in mammals, yeast, plants and C. elegans. Although it has not been defined an obvious REC8 homologue in Drosophila melanogaster, C(2)M is a protein distantly related to REC8 which has not play a cohesion role but localized to LEs and interacts with SMC3. In rat spermatocytes the meiosis-specific REC8 protein was observed as axial structures during premeiotic S phase, earlier than SMC1 β and SMC3 which appeared during leptotene along with SYCP2 and SYCP3, suggesting an special role of REC8 in AEs formation. The meiocytes of males and females Rec8 null mice do not complete the prophase I and presents SC assembly between sister chromatids. Thus, in mammals REC8 seems to be necessary to the formation of SC between homologous chromosomes instead of the properly assembly of SC, albeit its structure is altered. Recently a new meiosis-specific SCC1 cohesin subunit has been identified in vertebrates, it is closely related to RAD21 and it has been named RAD21L (RAD21-like). RAD21L interacts with other cohesin subunits such as SMC1 α/β, SMC3 or STAG3 but does not interact with REC8, thus both kleisins subunits form distinct meiosis-specific cohesin complexes.

There is no agreement regarding to the localization of RAD21L from mid-pachytene/ diplotene, thus meanwhile some authors ensure that RAD21L disappears from chromosomes axes at midpachytene when the SC is still formed, others have found that RAD21L is dissociated from AEs on autosomes in a progressive manner, during SC disassembly at diplotene and it is accumulated at centromeres until anaphase II. Similarly, there are different results in relation with the coexistence of RAD21 and RAD21L complexes during early prophase I stages. Lee and Hirano ensure that RAD21L seems to replace RAD21 from leptotene, finding that RAD21L localizes along AE/LEs until pachytene, accompanying the SC formation and homologous synapsis. However, RAD21 has been previously detected by immunolocalization during leptotene, zygotene and pachytene in mouse spermatocytes by other authors and at least in the first case, the antibody used in Parra's et al., study does not cross-react with RAD21L as it is shown in Herran's et al., paper., suggesting the presence of both RAD21 and RAD21L at the same time. At any case REC8, RAD21 and RAD21L are detected in a mutually exclusive manner, forming part of different cohesin complexes. The definitive evidence to ensure that RAD21L is directly implicated in AEs formation and synapsis is the Rad21L-/- male mouse which exhibits discontinuous AEs stretches and fails in homologous synapsis, moreover the mutants present defects in DSBs processing. The RAD21L knockout males are sterile but females are fertile although develop age-dependent sterility is observed. On the other hand, there are evidences of the interaction of RAD21L and SYCP1, main component of TFs of SC, and the centromere protein CENP-C. In a general manner, there is a rising common idea about the specific implication of RAD21L in pairing and the initiation of synapsis between homologous chromosomes. However, many questions are still open and new interesting doubts have been joined the fray regarding to the regulation of RAD21L cohesin complexes.

In mouse testis the meiotic protein STAG3 was first identified as a stromalin member associated to SC and after that was described its participation in the cohesin complex and its specificity to cohesion of sister chromatid arms during first meiotic division, however the authors pointed that STAG3 could be a component of SC, being detectable on chromatin paralleling to AEs appearance. This is an interesting example because STAG3 is the fourth component of the cohesin complex, it is not a kleisin and does not interact with SMC proteins which might be the base of the cohesin axes formed during the early stage of prophase I. Thus in a general model it has been observed a correlation between the progression of AE/LEs formation and synapsis and the localization of several cohesin subunits. According to the studied organisms the localization of cohesin subunits during prophase I has been observed continuous or discontinuous; however this could be related with the preparation method or the meiotic stage. In any case, all cohesin subunit are associated with AE/LEs during prophase I. What we consider more relevant is the sequential loading of cohesins observed during the first stages of prophase I and how the appearance of some subunits seems to be closely related with the progression of AE/LEs and SC formation and synapsis progression. In this sense, it is important to consider if the association of the cohesin subunits with the AE/LEs could be in an independent manner one to another. These cases would be possible only if we consider that in these species different cohesin complexes exist, as different author postulated before, being some of them more intimately related with the progression of synapsis or the SC formation. But then, we should think about if the cohesins loaded during early prophase I are really exerted a cohesin function or are related with AE/LEs formation and synapsis progression. Different analysis in mitotic cells showed that only the cohesins loaded during S phase are effective in the cohesion of sister chromatids, despite of evidences about new cohesion at DBSs point. Generally, the SMC cohesins seems to be more closely related with the AE/LEs formation, derivates from the evidences that this cohesin subunits are the base of the cohesin complexes then they would be the base of a cohesin axis and appear during the stages where SYCP2 and SYCP3 are incorporated or a bit early. However, the non-SMC proteins seem to have more specific roles before AEs formation or during SC assembly. Especially we are thinking in the SCC1 subunits and all its variants, and then a cohesin complex would acquire a specific role depending on the SCC1 which is formed with. It is difficult to split out the study of cohesin distribution during first meiotic division and the function of SC as the physical connector of homologous chromosomes at prophase I. More and more the scientists notice that the cohesins present function or contribute to function that historically had been attributed to the SC or to specific SC proteins and different cohesin complexes acquire specific functions in meiosis.

Cohesins and Meiotic Recombination

The first evidence that sister chromatid cohesion is required for DNA damage repair was the observation that the expression of mutated cohesins or the deletion of some subunits cause hypersensitivity and fails in DNA repair after induced or spontaneous DSBs in somatic line. Cohesin complexes are preferred loaded at intergenic sites onto the replicating chromosomes during S phase. However, now it is clear that in somatic cells the DNA damage induces the recruitment of cohesins after replication and lead to the formation of new cohesion, which is essential for DNA repair. Since the ability of SMC proteins to promote reannealing of complementary DNA strand was described, a role of cohesins in resolving DSBs was suggested. Different authors have studied and reviewed the role of cohesins in DSBs repair in S and G2 periods in mitotic cells, however, during meiosis the DSBs are programmed, in front of the spontaneous breaks that can alter the genome during the rest of the cell cycle. Basically mitotic/somatic and meiotic recombination have different purposes, thus whereas the meiotic recombination is initiated to generate a crossover (CO) on each bivalent at least, the mitotic recombination is used to repair DNA damage to ensure the fidelity of DNA sequence and cell surveillance. This important difference affects the repair pathway. We do not pretend to review all the literature regarding to the function of cohesin in all kinds of DNA damage, we will focus our attention in those cohesin subunits that have a specific role in meiotic DSBs and recombination.

Most of current models of meiotic recombination are based in the model that Szostak and colleagues proposed in yeast in 1983. This model is based on the generation of DSBs at early prophase I, the strand invasion at zygotene, the formation of double Holliday junction (dHJ) intermediate and its resolution at pachytene/diplotene. Obviously the repair mechanisms of DNA molecules act in an organized chromosome, where protein axial structures such as AEs and cohesin axis exist. The DSBs may be resolved toward CO between homologous chromatids or NCO (non-crossover), this decision can be made at different steps or repair pathway. Meiotic COs are essential for ensuring chromosome segregation. Chiasmata, points of genetic interchange between non-sister chromatids, keep the homologous chromosomes as a single structure after the disassembly of SC. Chiasmata and sister chromatid cohesion ensure the correct orientation of bivalents at metaphase I. Thus, in a similar way that fails in cohesion lead premature separation of chromatids, fails in meiotic recombination promote the generation of univalents at metaphase I. In both cases a high probability of aneuploidy exists. Meiotic recombination is initiated after a replication period, thus the sister chromatids are linked by cohesins when the recombination process begins. It has been established that cohesins are needed for the formation and stabilization of DNA interconnections.

Meiotic recombination begins via DSBs formation by the Topoisomerase II-like enzyme SPO11. As consequence of SPO11 activity a great amount of DSBs are generated during early meiotic prophase I and most of them are repaired as NCO. Thus the cohesins take part in bivalent stabilization and recombination to ensure that the excess of programmed DSBs do not alter the chromosome organization. On the other hand specific roles of different cohesin subunits in DSBs formation and regulation have been also described. As we stated above the basic difference between meiotic and mitotic recombination lies in the generation of DSBs, thus SPO11 has a pivotal role in the activation of meiotic recombination machinery and the regulation of this enzyme is one of the most interesting topics of meiotic recombination analysis. SPO11 has the key to begin all the process of genetic exchange between non-homologous chromosomes. Different studies have shown a non-random distribution of meiotic DSBs, giving raise to hot and cold-spots recombination sites. Different factors, as some histone variants or the DNA sequence, determine and regulate the action of SPO11. In this context the cohesins have also been implicated. In budding yeast the initial centromeric entry of SPO11 depends on REC8 and this subunit of cohesin complex choreographs the distribution of SPO11 to DSBs sites.

On the other hand, in S. pombe, the Rec8Δ mutant shows a marked reduction of meiosis-specific DNA breakage by REC12 (the S. pombe SPO11 homologue) and in C. elegans has been proposed that REC-8 could be implicated in limiting the activity of SPO-11. However, whereas in S. cerevisiae Rec8 mutants the chromosome segregation is at random, similar mutants in S. pombe present a equational segregation of sister chromatids at anaphase I. These discordant results suggest that the regulation of meiotic cohesin complexes containing REC8 and its role in recombination depends on the species. After the formation of DSBs by SPO11, modifications on chromatin as the phosphorylation of histone variants H2AX, H2Av and H2B occur and a cascade of repair proteins are loaded onto damage sites in order to restore the DNA properly. The first recombinases that bind to chromatin after DSBs are RAD51 and DMC1. The last one is a meiosis-specific protein. Usually the recruitment of recombinases has been associated to the AEs, but the cohesin axis is also functional in recruiting DNA recombination proteins. This issue was observed in Sycp3-/- mice spermatocytes, where the recombinase RAD51 is recruited in the absence of AEs suggesting a higher level of organization which is probably related to the cohesins. The resection of single-strand DNA allows the single-end invasion promoted by both recombinases RAD51 and DMC1. The extended invading strands are then processed to form CO or NCO. At this point the most of DBSs are resolved by Synthesis-Dependent Strand Annealing (SDSA) through NCO. Some points of strand invasion are stabilized to form a D loop and double Holliday junctions (dHJ) intermediate, which can be resolved through CO or NCO depending on the cleavage of dHJ.

The DSBs formation and the CO/NCO decision are two relevant items in meiotic recombination studies. In mice, before the AEs formation a cohesin core is observed as we reviewed above, but the action of SPO11 is not required for the formation of cohesin cores as it was observed in Spo11-/- spermatocytes, being SMC1$^{\beta}$ mostly associated to homologous or non-homologous synapsed regions meanwhile SMC3 localized both to AEs of unsynapsed and synapsed cores at arrested zygotene cells. In mammals, the absence of REC8 does not affect the timing of DSBs formation and the recruitment of the recombinases RAD51/DMC1, suggesting than DSBs and the early recombination events occur properly. But, the persistence of RAD51/DMC1 foci in Rec8-/- mice spermatocytes and the absence of MLH1 foci might reflect the inability to repair the DSBs. MLH1 and MLH3 protein are implicated in resolution of recombination to chiasma. It is not clear if REC8 has a specific role in resolving the meiotic DNA damage or the persistence of unresolved DSBs is a consequence of that the Rec8-/- cells do not reach the meiotic stage where the recombination resolution takes place. In this rodent REC8 seems to have a pivotal role in sister chromatid cohesion and synapsis but its relationship with recombination is not clear, despite of a small proportion of REC8 coprecipitate with RAD51/

DMC1. Moreover a small proportion of RAD50 coimmunoprecipitate with REC8 from spermatocyte lysates but REC8 does not coprecipitate with RAD50. RAD50, MRE11 and NBS1 proteins form the MRN complex which is implicated in the early response to DNA damage.

The action of MRN complex is related to nuclease activity at the 5' ends, releasing SPO11 from DNA. After the action of MRN, RAD51 and DMC1 can bind to single-strand DNA. These results which relate REC8 with recombination repair proteins would indicate that protein complexes that contain both REC8 and RAD51/DMC1 and some RAD50 exist in spermatocytes. The authors proposed that after S phase, cohesins would attract protein complexes that are involved in the early steps of homologous recombination. Recent results in yeast proposed that Rec8 could have a role in maintaining the homolog bias of recombination. In Rad21L-/- mice spermatocytes an accumulation of unrepaired DSBs has been described, but the cells are arrested at zygotenelike stage. Although, in sight of the cytological results of appearance and disappearance of RAD21L in wild type mice, a role in meiotic recombination has been suggested for this cohesin subunit, deeply studies are needed to corroborate the implication of RAD21L in the processing of recombination intermediates. However, in this context we might probably talk about a role of SMC1 α in recombination. Initially this protein was found at bridges between homologous AEs, and it was supposed that these represented the chiasmata sites. In Smc1β-/- spermatocytes early markers of recombination as RAD51 and γ-H2AX localize properly.

The cells progress through mid-pachytene and MLH1 and MLH3 foci were not detected in males, additionally the Smc1β-/- oocytes develop further and also fail to form the normal number of MLH1 foci. Probably the effects observing in cohesin mutants regarding to recombination proteins could be explain due to improper sister chromatid cohesion, an incorrect formation of AEs or synaptic fails. Perhaps we have to analyze the relationship between cohesin and meiotic recombination in a cohesin-dependent chromatin organization. An example is found in S. pombe, in this organism Rec8 and Rec11, meiosis-specific cohesin subunits, are essential for DSBs only in some regions of the genome. On the other hand the loading of Rec25 and Rec27, two recombination proteins in yeast, depends on the previous incorporation of Rec8. This issue could be explained because of Rec8 is required for the normal chromosome compaction during meiotic prophase I, thus it has been proposed that the regional effect on DSBs production might be related to the meiotic chromosomal organization formed by the cohesins. Thus the role of cohesins in meiotic recombination should be at different levels, chromosomal organization, providing the proximity between sister chromatid to ensure a correct repair with an undamaged template and as individual proteins that form complexes with recombination proteins.

Meiosis as an Evolutionary Adaptation for DNA Repair

The adaptive function of sex remains, today, one of the major unsolved problems in biology. Fundamental to achieving a resolution of this problem is gaining an understanding of the function of meiosis. The sexual cycle in eukaryotes has two key stages, meiosis and syngamy. In meiosis, typically a diploid cell gives rise to haploid cells. In syngamy (fertilization), typically two haploid gametes from different individuals fuse to generate a new diploid individual. A unique feature of meiosis, compared to mitosis, is recombination between non-sister homologous chromosomes. Usually these homologous chromosomes are derived from different individuals. In mitosis, recombination can occur, but it is ordinarily between sister homologs, the two products of a round of chromosome replication. Birdsell and Wills have reviewed the various hypotheses for the origin and maintenance of sex and meiotic recombination, including the hypothesis that sex is an adaptation for the repair of DNA damage and the masking of deleterious recessive alleles. Recently, we presented evidence that among microbial pathogens, sexual processes promote repair of DNA damage, especially when challenged by the oxidative defenses of their biologic hosts. Here, we present evidence that meiosis is primarily an evolutionary adaptation for DNA repair. Since our previous review of this topic, there has been a considerable increase in relevant information at the molecular level on the DNA repair functions of meiotic recombination.

Meiosis in protists and simple multicellular eukaryotes is induced in response to stressful conditions that likely cause DNA damage. Eukaryotes appeared in evolution more than 1.5 billion years ago. Among extant eukaryotes, meiosis and sexual reproduction are ubiquitous and appear to have been present early in eukaryote evolution. Malik et al. found that 27 of 29 tested meiotic genes were present in Trichomonas vaginalis, and 21 of these 29 genes were also present in Giardia intestinalis, indicating that most meiotic genes were present in a common ancestor of these species. Since these lineages are highly divergent among eukaryotes, these authors concluded that each of these meiotic genes were likely present in the common ancestor of all eukaryotes. Dacks and Roger also proposed that sex has a single evolutionary origin and was present in the last common ancestor of eukaryotes. Recently, this view received further support from a study of amoebae. Although amoebae generally have been assumed to be asexual, Lahr et al. showed that the majority of amoeboid lineages were likely anciently sexual, and that most asexual groups have probably arisen recently and independently.

Eukaryotes arose in evolution from prokaryotes, and eukaryotic meiosis may have arisen from bacterial transformation, a naturally occurring sexual process in prokaryotes. The fundamental similarities between transformation and meiosis have been explored. Bacterial transformation, like meiosis, involves alignment and recombination between non-sister homologous chromosomes (or parts of chromosomes) originating from different parents. Both during transformation and meiosis, homologs of the bacterial recA gene play a central role in the strand transfer reactions of recombination, indicating a mechanistic similarity. Also, bacterial transformation is induced by environmental stresses that are similar to those that induce meiosis in protists and simple multicellular eukaryotes, suggesting that there was continuity in the evolutionary transition from prokaryotic sex to eukaryotic sex. Evidence indicates that bacterial transformation is an adaptation for repairing DNA. Thus meiosis may have emerged from transformation as an adaptation for repairing DNA.

Among extant protists and simple multicellular eukaryotes sexual reproduction is ordinarily facultative. Meiosis and sex in these organisms is usually induced by stressful conditions. The paramecium tetrahymena can be induced to undergo conjugation leading to meiosis by washing, which causes rapid starvation. Depletion of the nitrogen source in the growth medium of the unicellular green alga Chlamydomonas reinhardi leads to differentiation of vegetative cells into gametes. These gametes can then mate, form zygotes and undergo meiosis. Upon nitrogen starvation or desiccation, the human fungal pathogen Cryptococcus neoformans undergoes mating or fruiting, both processes involving meiosis.

In addition to starvation, oxidative stress is another condition that induces meiosis and sex. The haploid fission yeast Schizosaccharomyces pombe is induced to undergo sexual development and mating when the supply of nutrients becomes limiting. Moreover, treatment of late-exponential-phase S. pombe vegetative cells with hydrogen peroxide, which causes oxidative stress, increases the frequency of mating and production of meiotic spores by 4- to 18-fold. The oomycete Phytophthora cinnamomi is induced to undergo sexual reproduction by exposure to the oxidizing agent hydrogen peroxide or mechanical damage to hyphae. In the simple multicellular green algae Volvox carteri, sex is induced by heat shock. This effect can be inhibited by antioxidants, indicating that the induction of sex by heat shock is mediated by oxidative stress. Furthermore, induction of oxidative stress by an inhibitor of the mitochondrial electron transport chain also induced sex in V. carteri. The budding yeast Saccharomyces cerevisiae reproduces as mitotically dividing diploid cells when nutrients are plentiful, but undergoes meiosis to form haploid spores when starved. When S. cerevisiae are starved, oxidative stress is increased and DNA double-strand breaks (DSBs) and apurinic/apyrimidinic sites accumulate. Perhaps, in S. cerevisiae, the induction of sex by starvation is mediated by oxidative stress, analogous to the way induction of sex by heat is mediated by oxidative stress in V. carteri.

These observations suggest that meiosis is an adaptation for dealing with stress, particularly oxidative stress. It is well established that oxidative stress induces a variety of DNA damages including DNA DSBs, single-strand breaks and modified bases. Thus we hypothesize that, in facultative sexual protists and simple multicellular eukaryotes, sex, with the central feature of meiosis, is an adaptive response to DNA damage, particularly oxidative DNA damage.

DNA Damages Induced by Exogenous Agents

If recombination during meiosis is an adaptation for repairing DNA damages, then it would be expected that exposure to DNA damaging treatments would increase the frequency of recombination, as measured by crossovers between allelic markers. Stimulation of allelic recombination was reported in the fruitfly Drosophila melanogaster in response to exposure to the DNA damaging agents UV light, X-rays, and mitomycin C. X-rays induce recombination in meiotic cells not only of D. melanogaster females, but also of males, which normally display no recombination during meiosis. Increased meiotic recombination in response to X-irradiation has also been reported in Caenorhabditis elegans, and in S. cerevisiae.

During Mitosis, Meiosis and DNA Damages caused by Diverse Exogenous Agents

Molecular recombination (that is homologous physical exchange or informational exchange) during mitosis and meiosis functions as a DNA repair process designated homologous recombinational repair (HRR). Many of the gene products employed in mitotic HRR are also employed in recombination during meiosis. It is this consistent function of recombination across meiosis and mitosis in eukaryotes and transformation in prokaryotes that we seek to understand through the repair hypothesis. Mutants defective in HRR genes in D. melanogaster and yeast have reduced ability to repair DNA damages arising from a variety of exogenous sources. These mutants are also defective in recombination during meiosis. In general, loss of HRR capability causes increased sensitivity to killing by agents that harm cells primarily through induction of DNA damage. These agents are listed in table. There have been no reports, that we know of, that HRR defective cells are sensitive to agents that harm cells by mechanisms other than primarily causing DNA damage.

In D. melanogaster, mutants defective in genes mei-41, mei-9, hdm, spnA and brca2 have reduced spontaneous allelic recombination (crossing over) during meiosis and increased sensitivity to killing by exposure to numerous DNA damaging agents. The Mei-41 protein is a structural and functional homolog of the human Atm (ataxia telangiectasia) protein, which plays a central role in HRR. The Mei-9 and Hdm proteins are components of a multiprotein complex that resolves meiotic recombination intermediates. The SpnA protein is a homolog of yeast Rad51, and Rad51 plays a central role in strand-exchange during HRR. The D. melanogaster Brca2 protein, a homolog of the human Brca2 protein that protects against breast cancer, regulates the activity of Rad51 protein in HRR. The Brca2 protein is required for HRR of DSBs during meiosis.

In S. cerevisiae, numerous mutant genes have been identified that confer sensitivity to radiation and genotoxic chemicals. Several of these mutant genes are also defective in meiotic recombination. For instance, the rad52 gene is required for meiotic recombination as well as for mitotic recombination. Mutants defective in the rad52 gene are sensitive to killing by several DNA damaging agents. Diploid cells of S. cerevisiae are able to repair DNA DSBs introduced by ionizing radiation, and this ability is lost in mutant strains defective in the rad52 gene. The Rad52 protein promotes the DNA strand exchange reaction of recombination during meiosis and mitosis.

Taken as a whole, these findings indicate that the products of genes mei-41, mei-9, hdm, spnA, and brca2 in D. melanogaster and the rad52 gene of yeast are required in meiosis for recombination and in somatic cells for HRR of potentially lethal DNA damages. Since the gene products that function in mitotic HRR are able to repair DNA damages from different sources, it can be reasonably assumed that these genes serve a similar DNA repair function during recombination in meiosis.

In the nematode C. elegans gonad, oocyte nuclei in the pachytene stage of meiosis, the stage in which HRR occurs, are hyper-resistant to X-ray irradiation compared to oocytes in the subsequent diakinesis stage of meiosis. This hyper-resistance depends on expression of gene ce-rdh-51, a homolog of yeast rad51 and dmc1 that play a central role in meiotic HRR. Meiotic pachytene nuclei are also more resistant to heavy ion particle irradiation than the subsequent meiotic diplotene or diakinesis stages. This resistance also depends on the ce-rdh-51 gene, as well as on gene ce-atl-1. ce-atl-1 is related to atm (ataxia –telangiectasia mutated), a gene necessary for repair of DSBs by HRR.

Coogan and Rosenblum measured repair of DSBs following γ-irradiation of rat spermatogenic cells during successive stages of germ cell formation. The stages were spermatagonia and preleptotene spermatocytes, pachytene spermatocytes and spermatid spermatocytes. The greatest repair capability was observed in pachytene, the stage of meiosis when HRR occurs. These findings indicate that HRR of γ-ray-induced DSBs occurs during meiosis. Several mammalian germ cell stages, including pachytene spermatocytes, produce levels of reactive oxygen species (ROS) sufficient to cause oxidative stress. This observation suggests that HRR during meiosis may also remove DNA damages caused by natural endogenously produced ROS.

The results reviewed indicate that, in both meiosis and mitosis, DNA damages caused by different exogenous agents are repaired by HRR, suggesting that DNA damages from natural endogenous sources (e.g. ROS) are similarly repaired. In general, DNA damage appears to be a fundamental problem for life. As noted by Haynes, DNA is composed of rather ordinary molecular subunits, which are not endowed with any peculiar kind of quantum mechanical stability. He observed that its very "chemical vulgarity" makes DNA subject to all the "chemical horrors" that might befall any such molecule in a warm aqueous medium. The average amount of oxidative DNA damage occurring per cell per day is estimated to be about 10,000 in humans, and in rat, with a higher metabolic rate, about 100,000. Most of these damages affect only one strand of the DNA, but a fraction, about 1-2%, are double-strand damages such as DSBs. These damages can be repaired accurately by HRR. In humans and rodents, defects in HRR enzymes lead to infertility, as would be expected if removal of DNA damages is an essential function of meiosis.

About 15% of all couples in the US are infertile, and an important cause of male infertility appears to be oxidative stress during gametogenesis. During spermatogenesis in the mouse, DNA repair capability declines after meiosis is complete, allowing accumulation of DNA damage. Lewis and Aitken reviewed evidence that DNA damages in the germ line of men are associated with poor semen quality, low fertilization rates, impaired pre-implantation development, increased abortion, and elevated incidence of disease in the offspring including childhood cancer. They noted that the natural causes of this DNA damage are uncertain, but the major candidate is oxidative stress. On the hypothesis that meiosis is an adaptation for DNA repair, it is expected that loss of ability to repair DNA damages during meiosis would have adverse effects, including infertility. Although the finding of such adverse effects is expected on the hypothesis that meiosis is an adaptation for repairing naturally caused DNA damages, this finding does not prove the hypothesis. Another possibility is that during meiosis damages are introduced in a programmed fashion, leading to HRR. Such HRR may be necessary for proper pairing and segregation of chromosomes, and this process may be required for fertility. Inherited mutations in genes that specify proteins necessary for HRR cause infertility indicating that production of functional gametes depends on HRR. Genes brca1, atm, and mlh1 are expressed in mitosis, but at a higher level in meiosis, and gene dmc1 is expressed exclusively in meiosis.

Brca1 functions during both meiotic and mitotic recombination. The inheritance of a mutant brca1 allele substantially increases a woman's lifetime risk for developing breast or ovarian cancer due to a deficiency in HRR of DNA DSBs in somatic cells. Male brca1 defective mice are infertile due to meiotic failure during spermatogenesis, indicating that HRR is necessary during meiosis.

The Atm protein acts during both meiotic and mitotic recombination in detection and signaling of DSBs, and is necessary for fertility of females and males in both humans and mice. Gametogenesis is severely disrupted in Atm-deficient mice as early as the leptonema stage of prophase I, resulting in apoptotic degeneration. Mismatch repair protein Mlh1 (homolog of E. coli MutL) is necessary for meiotic recombination. Mutation in the mlh1 gene causes blockage at the pachytene stage of meiosis and female and male infertility. Dmc1 is a meiosis specific gene. Dmc1 protein (a homolog of E. coli RecA protein) functions during meiotic recombination to promote recognition of homologous DNA and to catalyze strand exchange. Dmc1 deficient female and male mice are infertile due to arrest of gametes in meiotic prophase. The evidence reviewed indicates that defective HRR of DNA damages during meiosis causes infertility.

1. Non-crossover (Nco) Recombination During Meiosis is Likely an Adaptation for Dna Repair:

Meiotic recombination appears to be a near universal feature of meiosis [although it may be absent in some situations, such as in Drosophila males]. There are two major classes of meiotic recombination. If, during recombination, the chromosome arms on opposite sides of a DSB exchange partners, the recombination event is referred to as a crossover (CO). If the original configuration of chromosome arms is maintained, the recombination event is referred to as a non-crossover (NCO). The relative occurrence of NCO or CO recombination events is relevant to evolutionary theories of meiosis which assume producing genetic variation is the function of meiosis. NCO events have little effect on linkage disequilibrium (the statistical association of genes at different loci) and so produce very little genetic variation in terms of new combinations of genes. However, CO and NCO events are equivalent from the point of view of HRR. Data based on tetrad analysis from several species of fungi indicates that the majority (about 2/3) of recombination events during meiosis are NCOs. More recent work also supports a bias towards NCOs during meiosis. In mouse meiosis there are > 10-fold more DSBs than CO recombinants, suggesting that most DSBs are repaired by NCO recombination. In D. melanogaster there is at least a 3:1 ratio of NCOs to COs. These observations indicate that the majority of recombination events are NCOs. These NCOs involve informational exchange between two homologs but not physical exchange, and little genetic variation is created. Thus explanations for the adaptive function of meiosis that focus exclusively on crossing over are inadequate to explain the majority of recombination events. Andersen and Sekelsky have argued that a common mechanism called "synthesis dependent strand annealing" is employed in both meiotic HRR of the NCO type and mitotic HRR (which is largely of the NCO type), and thus meiotic and mitotic NCOs probably have a similar function. Substantial evidence indicates that HRR during mitosis is an adaptation to repair DNA damages that originate from diverse endogenous and exogenous sources (e.g. endogenous ROS from oxidative metabolism and exogenous Xrays, UV, chemical carcinogens). Thus NCO recombination during meiosis, as in mitosis, likely functions to repair of DNA damages from diverse sources.

2. NCO recombination likely occurs by synthesis-dependent strand annealing:

Molecular models of meiotic recombination have evolved over the years as relevant evidence accumulated. The model that has been most influential in recent decades has been the Double-Strand Break Repair model. By this model, during each recombination event two Holliday Junctions (HJs) are formed and resolved. Thus the Double-Strand Break Repair model can also be referred to as the Double Holliday Junction (DHJ) model. The DHJ model was considered to provide an explanation for both CO and NCO types of recombination events. However, Allers and Lichten showed that, although CO recombinants are likely formed by a pathway involving resolution of Holliday junctions, NCO recombinants arise by a different pathway that acts earlier in meiosis. Allers and Lichten, McMahill et al. and Andersen and Sekelsky have presented evidence that NCO recombinants are generated during meiosis by an HRR repair process referred to as "Synthesis-Dependent Strand Annealing" or "SDSA". During SDSA the invading strand from a chromosome with a DSB is displaced from the D-loop structure of an intact chromosome and its newly synthesized sequence anneals to the other side of the break on the chromosome with the original DSB. This process can accurately repair DNA DSBs by copying the information lost in the damaged homolog from the other intact homolog without the need for physical exchange of DNA. This process contributes little to genetic variation since the arms of the chromosomes flanking the recombination event remain in the parental position. Youds et al. presented evidence that the RTEL-1 protein of C. elegans physically dissociates strand invasion events, thereby promoting NCO repair by SDSA. HRR events initiated by DSBs consequently divide into two subsets, a larger subset which undergoes SDSA forming NCO recombinants, and a smaller subset which undergo DHJ repair and form CO recombinants. Perhaps SDSA is the preferred mode of HRR for unprogrammed double-strand damages, and DHJ repair is used primarily for programmed DSBs to promote proper chromosome segregation.

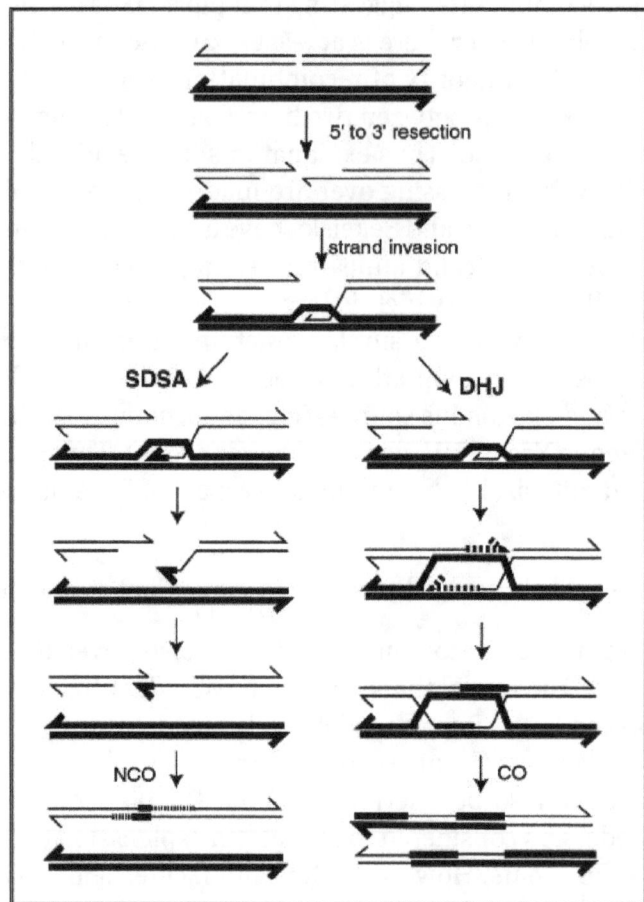

Current models of meiotic recombination are initiated by a double-strand break or gap, followed by pairing with an homologous chromosome and strand invasion to initiate the recombinational repair process. Repair of the gap can lead to crossover (CO) or non-crossover (NCO) of the flanking regions. CO recombination is thought to occur by the Double Holliday Junction (DHJ) model, illustrated on the right, above. NCO recombinants are thought to occur primarily by the Synthesis Dependent Strand Annealing (SDSA) model, illustrated on the left, above. Most recombination events appear to be the SDSA type.

Although the SDSA model starts with a DSB, it would also be applicable to other types of double-strand damages such as interstrand-crosslinks, or a single-strand damage (e.g. an altered base) opposite a break in the other strand. In principle, both of these types of doublestrand damages could be converted by nucleases to a DSB that would then be subject to SDSA.

3. The Role of Spo11 in Promoting Accurate DNA Repair can also Facilitate Proper Chromosome Segregation:

In the budding yeast S. cerevisiae, synapsis (pairing of homologous chromosomes) and synaptonemal complex formation depend on Spo11, a nuclease related to type II topoisomerases. Spo11 induces DSBs leading to HRR events of the CO type that form the physical association between homologs (chiasmata) needed for synaptonemal complex formation and proper disjunction of non-sister homologs at the first meiotic division. On the basis of these properties of Spo11, it is sometimes assumed that the primary function of meiotic recombination is to promote synapsis. However, as reviewed by Barzel and Kupiec, this theme cannot be generalized, as synapsis occurs independently of Spo11 induced recombination in the nematode worm C. elegans and the fruitfly D. melanogaster. In C. elegans, synapsis between homologs occurs normally in a spo-11 mutant. The D. melanogaster gene mei-W68 encodes a spo11 homolog. In D. melanogaster females, meiotic chromosome synapsis occurs in the absence of mei-W68 mediated CO recombination. Electron microscopy of oocytes from females homozygous for mei-W68 mutations that eliminated meiotic recombination revealed normal synaptonemal complex formation. In D. melanogaster females, meiotic recombination does not appear to be necessary for synapsis.

Since the role of Spo11 is of substantial interest in current discussions of the adaptive significance of meiotic recombination, we offer a speculation on its possible role consistent with the DNA repair hypothesis. both the DHJ and SDSA models for HRR start with a DSB. During meiosis in S. cerevisiae, DSBs are formed by a process that usually depends on Spo11. In S. pombe, Spo11 homolog Rec12 generates meiotic recombinants and meiosis specific DSBs. In C. elegans, a Spo11 homolog seems to have a similar role. We propose that DNA damages of various types are converted to DSBs, a "common currency," in order to initiate their recombinational repair. Spo11 appears to be employed in this process. Our reasoning is based on the precedents of the well-established pathways of nucleotide excision repair and base excision repair. In nucleotide excision repair, the initial steps of the pathway involve recognition of a wide variety of bulky damages followed by their removal to generate a single-strand gap, the "common currency" which is then repaired by a gap filling process. In base excision repair, a variety of altered bases are recognized by a corresponding variety of DNA glycosylases that generate an intermediate apurinic/apyrimidinic site, the "common currency" for further repair. On this reasoning, formation of DSBs by a Spo11-dependent process is part of an overall DNA repair sequence. In those species where the resolution of meiotic HRR by CO recombination is beneficial in promoting proper chromosome segregation at the first meiotic division, we think this benefit arose secondarily to the primary benefit of accurate DNA repair.

The function of recombination as a repair process may have arisen very early in the evolution of life [perhaps in the RNA world], and the function of promoting synapsis during meiosis probably arose later in evolution in some eukaryotic lineages. If, in mammals, a major function of meiotic CO recombination, as distinct from NCO recombination, is to promote synapsis and proper chromosome segregation, then one might expect CO events to be localized to specific hot-spot sequences. Hot-spot determinants may also include specific proteins that bind to hot-spot sequences and facilitate CO recombination such as Prdm9. It is estimated that, in humans, the average number of endogenous DNA DSBs per somatic cell occurring at each cell generation is about 50. This rate of DSB formation likely reflects unprogrammed damages, such as may be caused by ROS, and can be taken as an indication of the level of unprogrammed DSBs present in cells undergoing meiosis as well. In the human genome 25,000 hotspots for meiotic recombination have been identified. The average number of CO recombination events per hotspot is one CO event per 1,300 meioses. The large number of recombination hotspots is consistent with a wide distribution of sites vulnerable to unprogrammed DNA damage as well as specific sites where recombination would need to be induced to promote synapsis. A challenge for future research is the identification of the types of natural damages and programmed damages, and their frequencies, that are removed by CO recombinational repair during meiosis.

4. During meiosis, CO Recombination can Repair Dna Damages Independently of Spo11:

In a spo11 mutant of S. cerevisiae, the meiotic defects in recombination and synapsis are alleviated by X-irradiation, indicating that X-ray induced DNA damages can initiate CO recombination leading to synapsis independently of Spo11. Also, in C. elegans, Spo11 is required for meiotic recombination, but radiation induced-breaks alleviate this dependence. These findings indicate that unprogrammed DNA damages induced by X-rays can be repaired by HRR during meiosis independently of Spo11. In both S. pombe and C. elegans, mutants deficient for Spo11 undergo meiotic CO recombination when single base lesions of the type dU:dG are produced in their DNA. This recombination does not involve production of large numbers of DSBs, but does require uracil DNA-glycolylase, an enzyme that removes uracil from the DNA backbone and initiates base excision repair. These authors proposed that base excision repair of a uracil base, an abasic site, or a single-strand nick are sufficient to initiate meiotic CO recombination in S pombe and C. elegans. In a Rec12 (Spo11 homolog) mutant strain of S. pombe, meiotic recombination can be restored to near normal levels by a deletion in rad2 that encodes an endonuclease involved in Okazaki fragment processing. Both CO and NCO recombination were increased, but DSBs were undetectable. On the basis of the biochemical properties of Rad 2, these authors proposed that meiotic recombination can be initiated by non-DSB lesions, such as nicks and gaps, which accumulate during premeiotic DNA replication when Okasaki fragment processing is deficient.

In general the findings reviewed in indicate that DNA damages arising from a variety of sources can be repaired by meiotic HRR of the CO type, and that this repair may occur independently of Spo11. DNA repair likely provides the strong short-term advantage that maintains meiosis, while genetic variation may provide a long-term advantage.

Evolutionary explanations for sex have often assumed that the adaptive advantage of meiosis arises from the genetic variation produced. A variety of models and reviews have been presented in this active area of research. However, Otto and Gerstein have also pointed out that in a fairly stable environment, individuals surviving to reproductive age have genomes that function well in their current environment. They raise the question of why such individuals should risk shuffling their genes with those of another individual, as happens during meiotic recombination. This consideration, and others, have led many investigators to question whether production of genetic diversity is the principal adaptive advantage of sex. Heng and Gorelick and Heng reviewed evidence that sex actually decreases most genetic variation. Their view is that sex acts like a coarse filter, weeding out major changes, such as chromosomal rearrangements, but allowing minor variation, such as changes at the nucleotide or gene level (that are often neutral), to flow through the sexual sieve. Thus, they consider that sex acts as a constraint on genomic variation, thereby limiting adaptive evolution.

We consider that the major adaptive advantage of meiosis is enhanced recombinational repair. In contrast to the variation hypothesis, DNA repair provides an appropriate explanation for the adaptive advantage of sex (and meiosis) in the short-term, since its benefits are large enough (removal of DNA damages that would be deleterious/lethal to gametes or progeny) to plausibly balance the large costs of sex. The large costs of sex include the "cost of males", "recombinational load" that arises from the randomization of genetic information during sex and loss of coadapted gene complexes, the cost of mating, and cost of sexually transmitted disease.

The hypothesis that meiosis is an adaptation for DNA repair can be consistently applied to all organisms that have sex, including the facultative sexual organisms discussed above, as well as species that undergo meiosis but experience little or no outcrossing, as described below. If, in the long-term, the genetic variation produced by sex increases the rate of adaptation, as proposed by a number of authors, this would be an added benefit. However, in the short-term, we consider it unlikely that the benefit of variation is large enough to maintain sex.

In nature, many organisms that undergo meiosis outcross only rarely or not at all. In these cases, meiosis generates little or no genetic variation. In the budding yeast S. cerevisiae, outcrossing sex, in contrast to inbreeding sex, appears to be very infrequent in nature. Ruderfer et al. estimated that the ancestors of three S. cerevisiae strains outcrossed in nature only about once every 50,000 generations. On the other hand, mating between closely related yeast cells is likely to have been much more common in nature. Mating can occur when haploid cells of opposite mating types, MATa and MATα, come into contact. As pointed out by Zeyl and Otto, mating between closely related cells is common for two reasons: (1) The close physical proximity of cells of opposite mating type from the same ascus (the sac that contains the products from a single meiosis), and (2) Homothallism, the ability of haploid cells of one mating type to produce daughter cells of the opposite mating type. Thus, in nature, the meiotic events that produce little or no recombinational variation are much more frequent than meiotic events that do produce recombinational variation. This disparity is consistent with the idea that the primary adaptive function of meiosis in S. cerevisiae is HRR of DNA damages, since this benefit is realized in meiosis resulting from either inbreeding or outcrossing. If the primary adaptive function of meiosis were to generate genetic variation, it is difficult to understand how the complex process of meiosis could be selectively maintained in S. cerevisiae during the many generations in which there is no outcrossing.

Various levels of inbreeding due to consanguineous mating are known in many species. One extreme, but well studied, example among vertebrate species is the Mangrove Killifish, Kryptolebias marmoratus, which inhabits brackish water mangrove habitats from Brazil to Florida. These fish produce sperm and eggs by meiosis and reproduce routinely by selffertilization. Each hermaphroditic individual normally fertilizes itself when a sperm and egg that it has produced by an internal organ unite inside the fish's body. In this highly inbred hermaphroditic species meiotic recombination does not produce significant allelic variation, suggesting that meiosis is retained for some other adaptive benefit.

In higher plants, outcrossing sexual reproduction is the most common mode of reproduction, but about 15% of plants undergo meiosis and are principally self-fertilizing. We infer from these examples that the generation of genetic variation is not likely to be the adaptive benefit maintaining meiosis in these organisms. However, meiosis may be maintained by the adaptive benefit of HRR of DNA damage, since this benefit does not depend on outcrossing, nor that the participating chromosomes carry different alleles.

The meiotic function of repairing DNA damages primarily acts to preserve the existing genome. The generation of new genomic variants, a consequence of recombinational repair processes, appears to be a secondary effect that may provide a benefit in the long-term. As discussed above, most HRR events during meiosis are of the NCO type, which generate minimal genetic variation compared to the CO type. This is consistent with the DNA repair hypothesis, since both the CO and NCO types of recombination can repair DNA. On the assumption that the generation of variation is the primary benefit of meiosis, the majority of HRR events, those of the NCO type, provide no significant benefit and hence are wasteful. Even though, during meiosis, the frequency of CO recombination is ordinarily substantially less than the frequency of NCO recombination, during mitosis the frequency of CO compared to NCO recombination is even lower. The higher frequency of CO recombinants during meiosis compared to mitosis may reflect the role of CO recombinants in promoting synapsis during meiosis, a process distinct to meiosis.

5. During meiosis, hrr may remove a class of damages that cannot be accurately repaired during mitosis:

HRR during meiosis offers unique advantages compared to HRR during mitosis, based on the opportunity for non-sister homologs to pair and recombine during meiosis, which does not happen during mitosis. In mitosis, HRR involves interaction between the sisterchromosomes formed upon DNA replication. Thus, in mitosis, HRR is limited to the phases of the cell cycle during DNA replication (S phase) and after DNA replication (G2/M). Prior to DNA replication (G1 phase) in mitosis, double-strand DNA damages, such as DSBs, are repaired by an inaccurate process, non-homologous end-joining (NHEJ), which generates mutation. Double strand damages arising after DNA replication, may be repaired during mitosis by HRR between sisters. However, meiotic recombination can cope in a non-mutagenic way with double strand damages which arise at any point in the cell cycle.

Meiotic G1 phase cells appear to be more resistant to the lethal effects of X-irradiation than mitotic G1 phase cells. This finding suggests that repair of DSBs is more efficient during meiotic than mitotic G1 phase, as DSBs are a common consequence of Xirradiation. We speculate that during meiosis, in contrast to mitosis, double-strand damages occurring prior to DNA replication may be accurately repaired by HRR because pairing occurs between non-sister chromosomes. If this is so, meiotic cells have the advantage, compared to mitotic cells, of being able to accurately and efficiently repair double-strand damages that occur both before and after replication. As a result, germ cells would tend to be protected against the mutagenic effect of inaccurate NHEJ that typically occurs prior to replication in mitotic cells.

Mao et al. presented evidence that one type of somatic cell, human fibroblasts, utilizes error-prone NHEJ as the major DSB repair pathway at all cell cycle stages. In these cells, HRR is nearly absent prior to replication (G1 phase) and is used, when it occurs, primarily in the S phase. Even after the S phase when two sister-chromosomes are present (the G2/M phase), NHEJ is elevated and HRR is in decline.

The situation is somewhat different in mammalian embryonic stem (ES) cells compared to differentiated somatic cells. ES cells give rise to all of the cell types of an organism. Because mutations at this early embryonic stage are passed on to all clonal descendents, they can be seriously detrimental to the organism as a whole. Therefore robust mechanisms are needed in ES cells for reducing DNA damages (or eliminating damaged cells) in order to reduce mutations. Mouse ES cells were found to predominantly use high fidelity HRR to repair DSBs, compared to somatic cells that predominantly used NHEJ. Furthermore mouse ES cells lack a G1 checkpoint and do not undergo cell-cycle arrest upon receiving DNA damage prior to DNA replication. Rather, they undergo p53-independent apoptosis in response to DNA damage. Consistent with these findings, mouse ES stem cells have a mutation frequency about 100- fold lower than that of isogenic mouse somatic cells, but, as discussed next, at a likely cost resulting from somatic selection against cells with unrepairable DSBs which arise before DNA replication.

These results imply that a low mutation rate is achievable in mitotic cells by using apoptosis to remove cells with DNA damages that are present prior to replication, and using HRR, rather than NHEJ, to remove double-strand damages present subsequent to DNA replication. The non-sister chromosomes present in every diploid somatic cell during mitosis, in principal, might pair and undergo accurate HRR (as in meiosis), but this does not ordinarily occur, presumably because, in somatic cells, the benefit is outweighed by costs [e.g. loss of heterozygosity and expression of deleterious recessive alleles including those leading to cancer]. Meiosis is therefore unique, in that DNA damages occurring both prior to and after DNA replication can be subject to high fidelity HRR between non-sister homologs. This would avoid the high costs of both deleterious mutation and loss of potential gametes due to apoptosis.

In humans at each cell division, 30,000-50,000 DNA replication origins are activated. Thus the chromosome is ordinarily replicated in segments. We postulate that any segment containing a DSB will fail to complete its replication until the DSB is repaired. This limited and temporary blockage of replication may result directly from the break itself, or occur as a response to regulatory events set off by proteins that specifically bind to the broken ends. In any case, HRR can be carried out during the subsequent prophase I stage of meiosis, when the segment containing a DSB pairs with a non-sister homologue. This repair would then allow chromosome replication to be completed.

6. DNA Damage During the Mitotic Divisions of the Germ Line in Multicellular Organisms:

In multicellular eukaryotes there are typically many mitoses during germ line development, and only a single final meiosis leading to gamete formation. During the mitotic cell divisions in the germ line, DSBs and other double-strand damages occurring after DNA replication are likely repaired by HRR or eliminated from the cell lineage by death and apoptosis of the damaged cell. We have argued above that because of the lack of pairing of non-sister homologs during mitosis, HRR is unable to accurately repair double-strand damages occurring before replication. Thus when double-strand damages occur prior to replication during the mitotic divisions in the germ line the consequence will be either increased mutation or increased apoptosis. By analogy with the strategy used by somatic stem cells, we think that the preferred strategy during these mitotic divisions is likely to be apoptosis, since this avoids mutations in the germ line that could be passed on to progeny. However, double-strand damages occurring prior to replication during meiosis need not lead to apoptosis (which would likely decrease fecundity), since these can be accurately repaired by HRR between non-sister chromosomes. The consequence will be enhanced gamete viability and fecundity, that is, enhanced fitness. In the mitotic divisions of the germ-line prior to meiosis, loss of cells due to DNA damageinduced apoptosis need not be very costly to organism fitness, since such losses could be made up by extra cell divisions of undamaged cells. However, the loss of sperm or egg cells due to unrepaired DNA damage would likely have substantial costs to fitness due to loss of fertility and progeny.

Why is Meiosis Frequently Associated with Outcrossing?

While the focus of this article is on the adaptive benefit of meiosis itself, we briefly consider why meiosis is frequently associated with outcrossing, where the chromosomes involved in recombination come from different unrelated parents in a prior generation. Previously, we discussed examples of meiosis occurring in association with inbreeding and self-fertilization. Meiosis with inbreeding will be favored when the costs of mating are high (e.g. the cost of finding a mate at low population density). These examples of inbred meiosis were presented to illustrate our argument that meiosis provides an adaptive advantage (accurate DNA repair) independent of whether significant recombinational variation is also produced. However, meiosis is often associated with outcrossing, and we now consider why.

A disadvantage of inbreeding, especially of self-fertilization, is expression of deleterious recessive mutations, resulting in inbreeding depression. Analysis of the effects of masking deleterious recessive mutations (genetic complementation) using heuristic modes and arguments indicated that complementation provides benefits sufficient to maintain outcrossing. However, more explicit population genetic models have raised some issues that are in need of further clarification. In population genetics terms, the basic effect of outcrossing is to bring populations to Hardy-Weinberg (HW) equilibrium. Thus, outcrossing can be beneficial if there is another force that pushes the population away from HW equilibrium (generating either an excess or a deficit of heterozygotes) and if it's advantageous to go closer to HW equilibrium. One possible force that generates departure from HW equilibrium is dominance: for example if deleterious alleles tend to be recessive, after selection there will be an excess of heterozygotes (and a deficit of homozygotes).

However in this case outcrossing is costly in the short term (because it tends to expose deleterious alleles), but beneficial in the long term (because purging them becomes more efficient). Otto showed that under this scenario high rates of outcrossing are favored only if deleterious alleles are weakly recessive. Another potential force pushing away from HW equilibrium considered by Roze and Michod is gene conversion which creates homozygosity. Gene conversion could result from mitotic HRR between sister chromosomes as discussed above. In this case (and if deleterious alleles tend to be partially recessive) outcrossing is beneficial in the short term (because it masks deleterious alleles) but disadvantageous in the long term (because purging is less efficient). The magnitude of this force may be estimated from rates of loss of heterozygosity during development [discussed in Roze and Michod]. The few estimates which exist indicate that the loss of heterozygosity is low, and thus this selective force for outcrossing may be weak. Clearly, we need more estimates of this critical parameter to know how large this force for outcrossing may be. Another consequence of outcrossing is the generation of new genetic variants which may provide an additional long-term advantage.

Special Case of Asexual Bdelloid Rotifers

Bdelloid rotifers are common invertebrate animals. They are apparently obligate asexuals that reproduce by parthenogenesis. These organisms are extraordinarily resistant to ionizing radiation. This resistance appears to be a consequence of an evolutionary adaptation to survive desiccation in ephemerally aquatic habitats. Such desiccation causes extensive DNA breakage, which they are able to repair. Bdelloid primary oocytes are in the G1 phase of the cell cycle and thus lack sister chromatids. Welch et al. proposed a mechanism of repair involving interaction of non-sister co-linear chromosome pairs, which are maintained as templates for repair of DNA DSBs caused by the frequent desiccation and rehydration. Thus although these organisms apparently lack sex and meiosis, an essential feature of meiosis, HRR between non-sister homologs appears to be retained.

7. Conservation among eukaryotes of reca-like proteins as key components of the hrr machinery acting during meiosis:

Sex appears to be universally based on RecA-like proteins. RecA-like proteins play a key role in HRR, and the HRR machinery and its mechanism of action appear to be highly conserved among eukaryotes. The rad51 and dmc1 genes in the eukaryotic yeasts S. cerevisiae and S. pombe are orthologs of the bacterial recA gene. The dmc1 gene is found in many different eukaryote species, and has been reported, for instance, in the protists Giardia, Trypanosoma, Leishmania, Entamoeba and Plasmodium. Rad51 and Dmc1 proteins are recombinases that interact with single-stranded DNA to form filamentous intermediates called presynaptic filaments, and these filaments initiate HRR. Dmc1 recombinase functions only during meiosis, whereas Rad51 recombinase acts in both somatic HRR and in meiosis. When it functions in meiosis, Rad51 mainly uses a sister chromosome for HRR. In contrast, Dmc1 mainly uses the non-sister homologous chromosome. The yeast Rad51 recombinase catalyzes ATP-dependent homologous DNA pairing and strand exchange, as does the bacterial RecA recombinase. The tertiary structure of the Dmc1 recombinase has an overall similarity to the bacterial RecA recombinase. These observations suggest that the bacterial RecA that functions in the bacterial sexual process of transformation, and the yeast Rad51 and Dmc1 recombinases that act in meiosis have similar functions, consistent with the idea that meiotic recombination evolved from simpler sexual processes in bacteria.

We next consider evidence that RecA orthologs play a key role in meiosis, not only in protists, but also in multicellular eukaryotes. RecA orthologs act in meiosis in a range of animals (e.g. nematodes, chickens, humans and mice) and plants (e.g. Arabidopsis, rice and lilies). The rad51 gene is expressed at a high level in mouse testis and ovary, suggesting that Rad51 protein is involved in meiotic recombination. In mice, mutations in the dmc1 gene cause sterility, failure to undergo intimate pairing of homologous chromosomes and an inability to complete meiosis. In the nematode C. elegans, resistance to DNA damage caused by X-irradiation in the meiotic pachytene nuclei depends on a RecA-like gene. RecA gene orthologs are also expressed in chicken testis and ovary and in human testis. In humans, Dmc1, the meiosis-specific recombinase, forms nucleoprotein complexes on single-stranded DNA that promote a search for homology and carry out strand exchange, the two necessary steps of genetic recombination.

In lily plants, genes lim15 and rad51 are orthologs, respectively, of the dmc1 and rad51 genes of yeast. The lily proteins Lim15 and Rad51 colocalize on chromosomes in various stages of meiotic prophase I, and form discrete foci. The proteins of these foci are considered to participate in the search for, and pairing of, homologous sequences of DNA. In another plant, Arabidopsis thaliana, meiotic recombination requires Dmc1 and Rad51. In the rice plant, an ortholog of dmc1 is necessary for meiosis and has a key function in the pairing of homologous chromosomes.

In general, both animals and plants have RecA-like proteins that appear to have a central function in meiotic HRR. Furthermore, bacterial RecA and its animal and plant orthologs have very similar roles in the HRR events during the sexual processes of bacterial transformation and eukaryotic meiosis. In all cases, the RecA protein or RecA-like protein assembles on single-stranded DNA to form a pre-synaptic filament. This filament then attaches to a duplex DNA molecule and searches for homology in its target. When the presynaptic molecule locates an homologous sequence in the duplex molecule, it is able to form a DNA joint. These joints are then processed further to complete the HRR event.

MITOSIS VS. MEIOSIS

Cell Division

- Mitosis: A somatic cell divides once. Cytokinesis (the division of the cytoplasm) occurs at the end of telophase.

- Meiosis: A reproductive cell divides twice. Cytokinesis happens at the end of telophase I and telophase II.

Daughter Cell Number

- Mitosis: Two daughter cells are produced. Each cell is diploid containing the same number of chromosomes.

- Meiosis: Four daughter cells are produced. Each cell is haploid containing one-half the number of chromosomes as the original cell.

Genetic Composition

- Mitosis: The resulting daughter cells in mitosis are genetic clones (they are genetically identical). No recombination or crossing over occur.

- Meiosis: The resulting daughter cells contain different combinations of genes. Genetic recombination occurs as a result of the random segregation of homologous chromosomes into different cells and by the process of crossing over (transfer of genes between homologous chromosomes).

Length of Prophase

- Mitosis: During the first mitotic stage, known as prophase, chromatin condenses into discrete chromosomes, the nuclear envelope breaks down, and spindle fibers form at opposite poles of the cell. A cell spends less time in prophase of mitosis than a cell in prophase I of meiosis.

- Meiosis: Prophase I consists of five stages and lasts longer than prophase of mitosis. The five stages of meiotic prophase I are leptotene, zygotene, pachytene, diplotene, and diakinesis. These five stages do not occur in mitosis. Genetic recombination and crossing over take place during prophase I.

Tetrad Formation

- Mitosis: Tetrad formation does not occur.

- Meiosis: In prophase I, pairs of homologous chromosomes line up closely together forming what is called a tetrad. A tetrad consists of four chromatids (two sets of sister chromatids).

Chromosome Alignment in Metaphase

- Mitosis: Sister chromatids (duplicated chromosome comprised of two identical chromosomes connected at the centromere region) align at the metaphase plate (a plane that is equally distant from the two cell poles).

- Meiosis: Tetrads (homologous chromosome pairs) align at the metaphase plate in metaphase I.

Chromosome Separation

- Mitosis: During anaphase, sister chromatids separate and begin migrating centromere first toward opposite poles of the cell. A separated sister chromatid becomes known as daughter chromosome and is considered a full chromosome.

- Meiosis: Homologous chromosomes migrate toward opposite poles of the cell during anaphase I. Sister chromatids do not separate in anaphase I.

Similarities between Mitosis and Meiosis

Mitosis and meiosis have different purposes, but share common features in how they work. Knowing their similarities is the beginning of understanding how they are different. The fundamental difference between mitosis and meiosis is that mitosis produces two daughter cells with the same number of chromosomes as the parent cell. Meiosis results in four daughter cells harboring only half of their parent's chromosomes, which underwent recombination. Aside from these two distinct purposes, both mitosis and meiosis occur in multiple stages during which the same general things happen: DNA replication and condensation, nuclear membrane degradation, spindle formation, chromosomal segregation and nuclear reformation. The same mechanisms of chromosomal segregation are at work in both mitosis and meiosis -- centrosomes, microtubules and motor proteins. Both mitosis and meiosis are multistage processes. The stages are interphase, prophase, metaphase, anaphase and telophase. The same general processes occur in each of these stages for mitosis and meiosis. Interphase is cell growth and DNA replication in preparation for cell division. Prophase is when the nuclear membrane degrades. Metaphase is when the chromosomes align in the middle of the cell. Anaphase is when the chromosomes are pulled apart. Lastly, telophase is when one cell splits into two separate cells.

DNA Replication

Mitosis and meiosis both involve duplication of a cell's DNA content. Each strand of DNA, or chromosome, is replicated and remains joined, resulting in two sister chromatids for each chromosome. A common goal of mitosis and meiosis is to split the nucleus and its DNA content between two daughter cells. Before DNA can be separated between two cells, it must be duplicated and then condensed into a form that can be efficiently transported. The condensed form of DNA is called chromosomes. Chromosomes can be physically pulled apart, and make separating long strands of DNA easier for the cell because the DNA strands become like a tightly wrapped yarn ball.

The Same Road

The mechanism by which chromosomes are pulled apart is the same for mitosis as it is for meiosis. Two centrosomes separate into opposite corners of a dividing cell. The centrosomes produce long micro-tubules, which are like railroads upon which motor proteins can walk. These micro-tubules connect the centrosome, which is like a train station, to the chromosomes. During metaphase, the chromosomes are aligned in the middle of the cell. Motor proteins on the chromosomes grab on to the micro-tubules that connect the centrosome to the chromosome. The motors proteins walk towards the centrosome, pulling the sister chromatids apart, each moving towards opposite poles of the dividing cell.

Rebuilding the Roof

One of the essential functions of mitosis and meiosis is the duplication and separation of the cell's DNA content. This process is closely tied to the degradation and reformation of the nuclear membrane. Eukaryotes are organisms whose cells contain membrane-bound structures called organelles, one of which is the nucleus. The separation of chromosomes in mitosis and meiosis must be preceded by a degradation of the nuclear membrane so that the chromosomes can align in the middle of the cell and then be evenly split in two directions. After anaphase, when the chromosomes are pulled apart, the nuclear membrane must be reformed. The cell must restore the nuclear membrane to ensure the proper function of DNA after mitosis or meiosis is complete.

GENETIC VARIATION IN MITOSIS AND MEIOSIS

Diploid (2n)

2 sets of homologous chromosomes.

Haploid (1n)

1 single set of homologous chromosomes. This cell has 2 pairs of chromosomes; 1 long, 1 short.

There are two sets of 2 similar (homologous) chromosomes:

- Ploidy = diploid, 2n:

    ◦   1 of chromosomes = 4.

    ◦   2 of chromatids = 4.

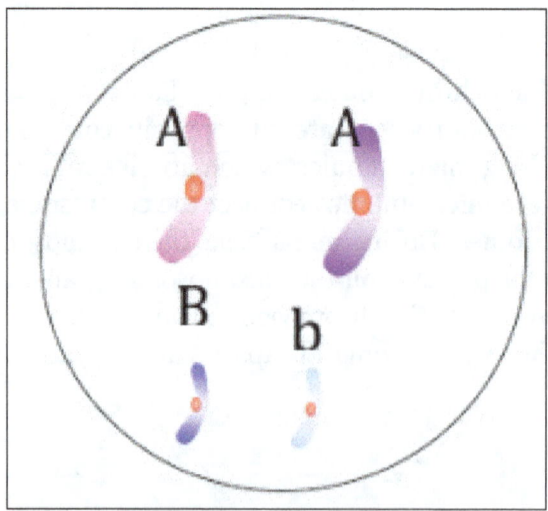

This cell has 2 pairs of duplicated homologous chromosomes; 1 long, 1 short.

- Ploidy = diploid, 2n:

    ◦   1 of chromosomes = 4.

    ◦   2 of chromatids = 8.

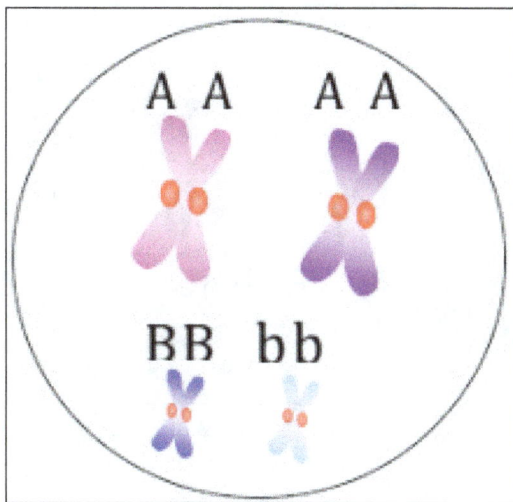

This cell has 2 chromosomes; 1 long, 1 short. There is only 1 copy of each chromosome, so it is haploid.

- Ploidy = haploid, 1n:

  - 1 of chromosomes = 2.

  - 2 of chromatids = 2.

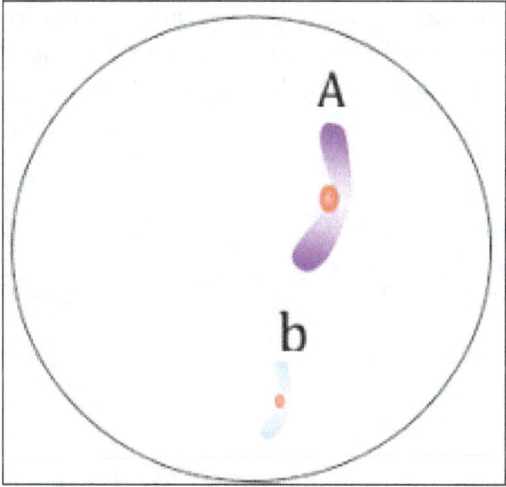

This cell has 2 duplicated chromosomes; 1 long, 1 short. There is only 1 copy of each chromosome, so it is haploid.

- Ploidy = haploid, 1n:

  - 1 of chromosomes = 2.

  - 2 of chromatids = 4.

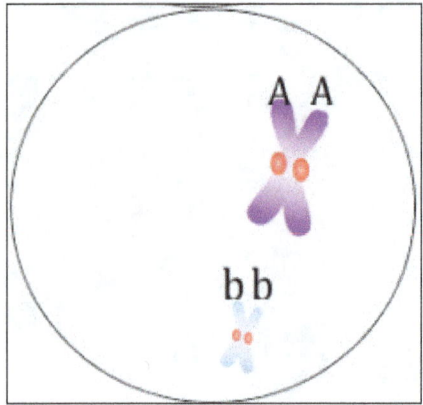

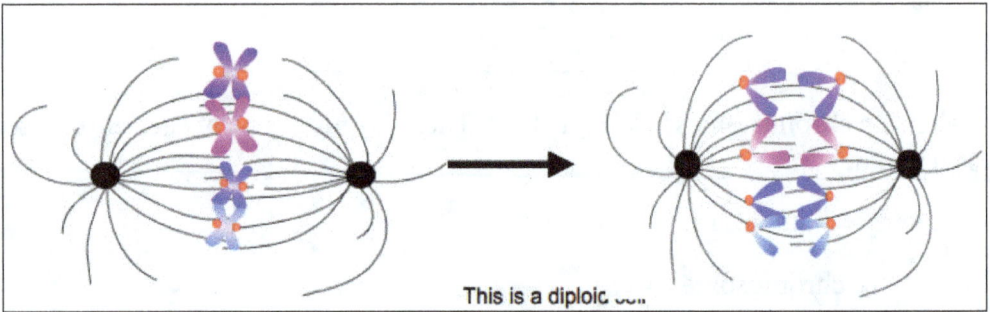

This is a diploid cell in metaphase and anaphase of mitosis. In the metaphase cell, there are 4 chromosomes (8 chromatids) total and two sets of homologous chromosomes that are duplicated. In the anaphase cell, there are 8 chromosomes. The resulting daughter cells will also be diploid and genetically identical to the mother cell.

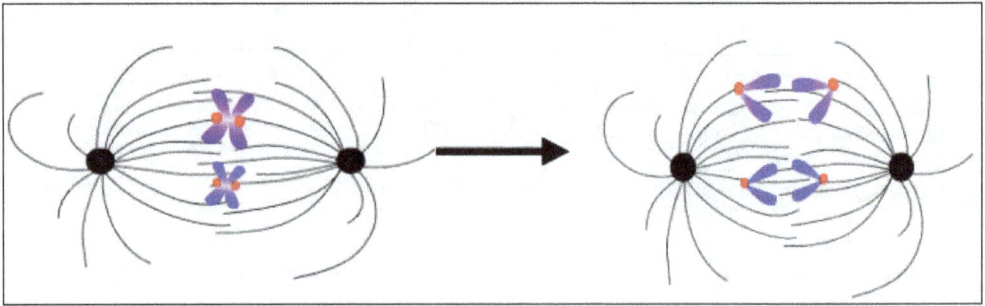

This is a haploid cell in metaphase and anaphase of mitosis. There are 2 chromosomes (4 chromatids); 1 big chromosome, 1 small chromosome in the metaphase cell. In the anaphase cell, there are 4 chromosomes present. The resulting daughter cells would be halploid and genetically identical to the mother cell.

There are two divisions in meiosis. The cell entering meiosis is diploid. In meiosis homologous chromosomes pair (allows crossing over of genetic material), but homologous do not pair in mitosis. Another difference is that after the first meiotic division, the cells do not reenter interphase and DNA is not replicated.

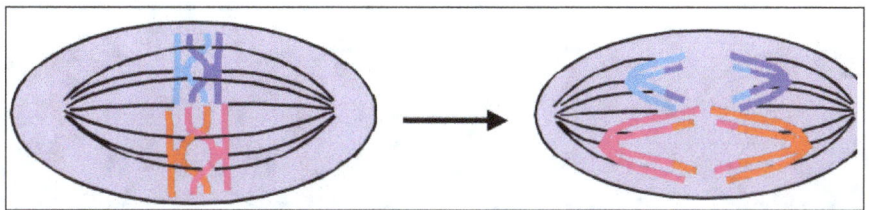

In metaphase 1 of meiosis, bivalents orient at the metaphase plate and homologous are paired. Each chromosome of a homologous pair attaches to fibers from opposite poles. The sister chromatids attach to fibers from the same pole. In anaphase 1, the centromere does not divide and homologous chromosomes move to opposite poles.

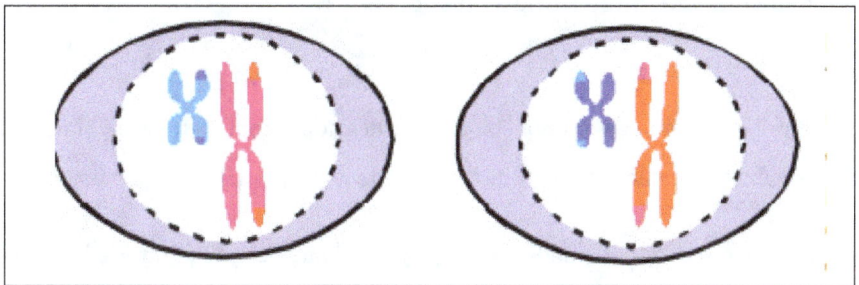

This is the separation of homologous chromsomes. These are the products of the first meiotic division. Only 1 of each chromosome (long and short) is present, therefore the daughter cells produced from the first meiotic division are haploid. In between meiosis 1 and 2, the DNA does not replicate and the starting cells are haploid.

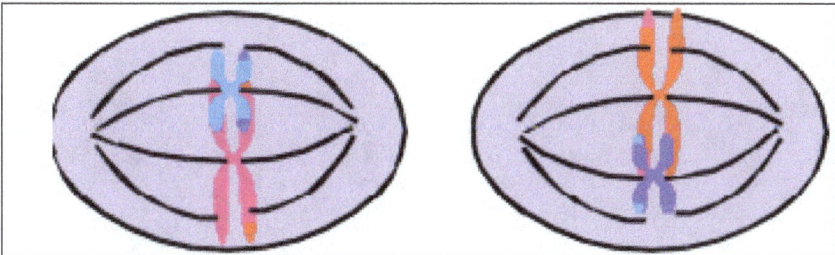

In metaphase 2, the chromosomes align at the metaphase plate and sister chromatids attach to spindle fibers from opposite poles.

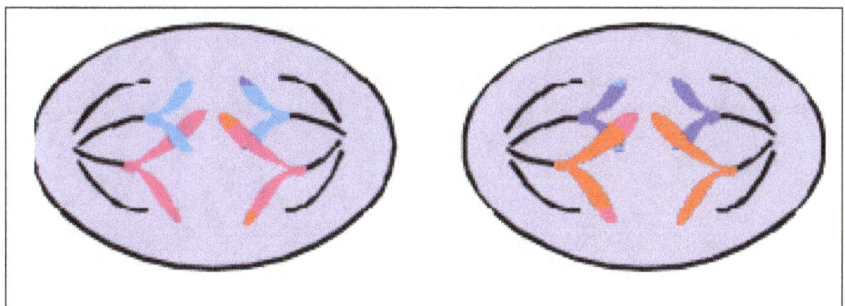

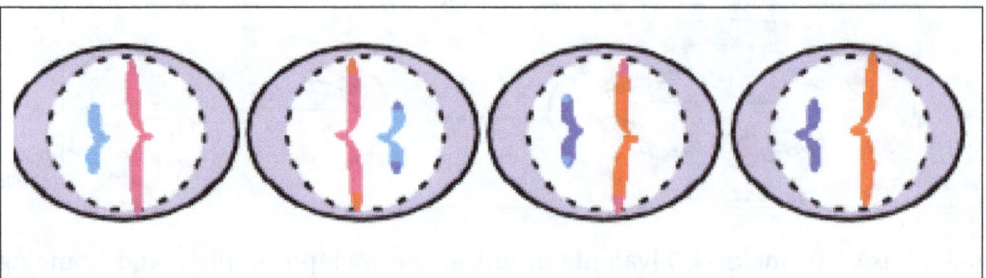

In anaphase 2, the centromeres divide and chromatids move to opposite poles. This is the separation of sister chromatids. The daughter cells produced are also haploid, having only 1 of each chromosome (long and short).

References

- Mitosis, cell-cycle-and-cell-division, biology, guides: toppr.com, Retrieved 1 June, 2020
- Mitosis-definition-features-and-significance-cell-division-37741: biologydiscussion.com, Retrieved 13 April, 2020
- Differences-between-mitosis-and-meiosis-373390: thoughtco.com, Retrieved 26 March, 2020
- Similarities-mitosis-meiosis-3866: education.seattlepi.com, Retrieved 30 May, 2020
- What-is-mitosis, facts: yourgenome.org, Retrieved 11 February, 2020
- Mitotic-inhibitors, drug-class: drugs.com, Retrieved 13 August, 2020

2

Cell Cycle, Division and Differentiation

A series of events that takes place in a cell as it grows and divides is termed as a cell cycle. The process in which a parent cell is divided into two or more daughter cells is called cell division. Cell differentiation refers to the process in which the dividing cells alter their functional type. This chapter discusses cell cycle, cell division and cell differentiation in detail.

CELL

Cells are considered the basic units of life in part because they come in discrete and easily recognizable packages. That's because all cells are surrounded by a structure called the cell membrane — which, much like the walls of a house, serves as a clear boundary between the cell's internal and external environments. The cell membrane is sometimes also referred to as the plasma membrane.

Cell membranes are based on a framework of fat-based molecules called phospholipids, which physically prevent water-loving, or hydrophilic, substances from entering or escaping the cell. These membranes are also studded with proteins that serve various functions. Some of these proteins act as gatekeepers, determining what substances can and cannot cross the membrane. Others function as markers, identifying the cell as part of the same organism or as foreign. Still others work like fasteners, binding cells together so they can function as a unit. Yet other membrane proteins serve as communicators, sending and receiving signals from neighboring cells and the environment — whether friendly or alarming.

Within this membrane, a cell's interior environment is water based. Called cytoplasm, this liquid environment is packed full of cellular machinery and structural elements. In fact, the concentrations of proteins inside a cell far outnumber those on the outside — whether the outside is ocean water (as in the case of a single-celled alga) or blood serum (as in the case of a red blood cell). Although cell membranes form natural barriers in watery environments, a cell must nonetheless expend quite a bit of energy to maintain the high concentrations of intracellular constituents necessary for its survival. Indeed, cells may use as much as 30 percent of their energy just to maintain the composition of their cytoplasm.

Other Components of Cells

A cell's cytoplasm is home to numerous functional and structural elements. These elements exist in the form of molecules and organelles — picture them as the tools, appliances, and inner rooms of the cell. Major classes of intracellular organic molecules include nucleic acids, proteins, carbohydrates, and lipids, all of which are essential to the cell's functions.

Nucleic acids are the molecules that contain and help express a cell's genetic code. There are two major classes of nucleic acids: deoxyribonucleic acid (DNA) and ribonucleic acid (RNA). DNA is the molecule that contains all of the information required to build and maintain the cell; RNA has several roles associated with expression of the information stored in DNA. Of course, nucleic acids alone aren't responsible for the preservation and expression of genetic material: Cells also use proteins to help replicate the genome and accomplish the profound structural changes that underlie cell division.

Proteins are a second type of intracellular organic molecule. These substances are made from chains of smaller molecules called amino acids, and they serve a variety of functions in the cell, both catalytic and structural. For example, proteins called enzymes convert cellular molecules (whether proteins, carbohydrates, lipids, or nucleic acids) into other forms that might help a cell meet its energy needs, build support structures, or pump out wastes.

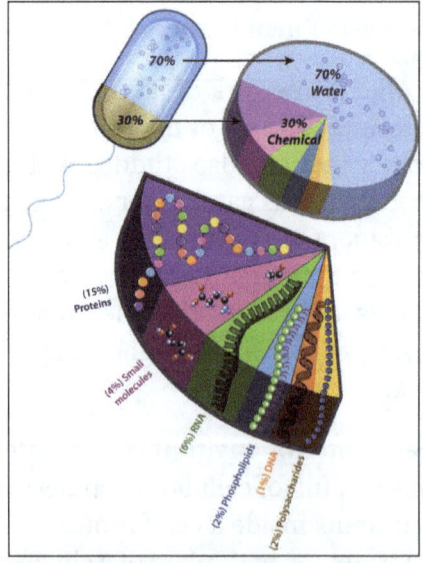

The composition of a bacterial cell.

Carbohydrates, the starches and sugars in cells, are another important type of organic molecule. Simple carbohydrates are used for the cell's immediate energy demands, whereas complex carbohydrates serve as intracellular energy stores. Complex carbohydrates are also found on a cell's surface, where they play a crucial role in cell recognition.

Finally, lipids or fat molecules are components of cell membranes — both the plasma membrane and various intracellular membranes. They are also involved in energy storage, as well as relaying signals within cells and from the bloodstream to a cell's interior).

Some cells also feature orderly arrangements of molecules called organelles. Similar to the rooms in a house, these structures are partitioned off from the rest of a cell's interior by their own intracellular membrane. Organelles contain highly technical equipment required for specific jobs within the cell. One example is the mitochondrion — commonly known as the cell's "power plant" — which is the organelle that holds and maintains the machinery involved in energy-producing chemical reactions. Most of a cell is water (70%). The remaining 30% contains varying proportions of structural and functional molecules.

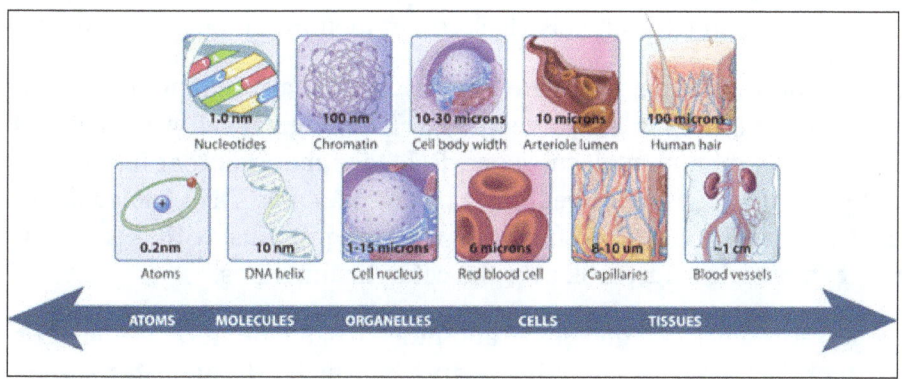

The relative scale of biological molecules and structures.

Cells can vary between 1 micrometer (µm) and hundreds of micrometers in diameter. Within a cell, a DNA double helix is approximately 10 nanometers (nm) wide, whereas the cellular organelle called a nucleus that encloses this DNA can be approximately 1000 times bigger (about 10 µm). See how cells compare along a relative scale axis with other molecules, tissues, and biological structures (blue arrow at bottom). Note that a micrometer (µm) is also known as a micron.

Different Categories of Cells

Rather than grouping cells by their size or shape, scientists typically categorize them by how their genetic material is packaged. If the DNA within a cell is not separated from the cytoplasm, then that cell is a prokaryote. All known prokaryotes, such as bacteria and archaea, are single cells. In contrast, if the DNA is partitioned off in its own membrane-bound room called the nucleus, then that cell is a eukaryote. Some eukaryotes, like amoebae, are free-living, single-celled entities. Other eukaryotic cells are part of multicellular organisms. For instance, all plants and animals are made of eukaryotic cells — sometimes even trillions of them.

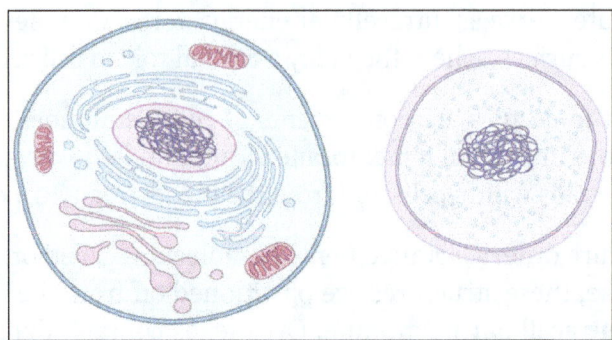

Comparing basic eukaryotic and prokaryotic differences.

A eukaryotic cell (left) has membrane-enclosed DNA, which forms a structure called the nucleus (located at center of the eukaryotic cell; note the purple DNA enclosed in the pink nucleus). A typical eukaryotic cell also has additional membrane-bound organelles of varying shapes and sizes. In contrast, a prokaryotic cell (right) does not have membrane-bound DNA and also lacks other membrane-bound organelles as well.

Researchers hypothesize that all organisms on Earth today originated from a single cell that existed some 3.5 to 3.8 billion years ago. This original cell was likely little more than a sac of small organic molecules and RNA-like material that had both informational and catalytic functions. Over time, the more stable DNA molecule evolved to take over the information storage function, whereas proteins, with a greater variety of structures than nucleic acids, took over the catalytic functions.

The absence or presence of a nucleus — and indeed, of all membrane-bound organelles — is important enough to be a defining feature by which cells are categorized as either prokaryotes or eukaryotes. Scientists believe that the appearance of self-contained nuclei and other organelles represents a major advance in the evolution of cells. But where did these structures come from? More than one billion years ago, some cells "ate" by engulfing objects that floated in the liquid environment in which they existed. Then, according to some theories of cellular evolution, one of the early eukaryotic cells engulfed a prokaryote, and together the two cells formed a symbiotic relationship. In particular, the engulfed cell began to function as an organelle within the larger eukaryotic cell that consumed it. Both chloroplasts and mitochondria, which exist in modern eukaryotic cells and still retain their own genomes, are thought to have arisen in this manner.

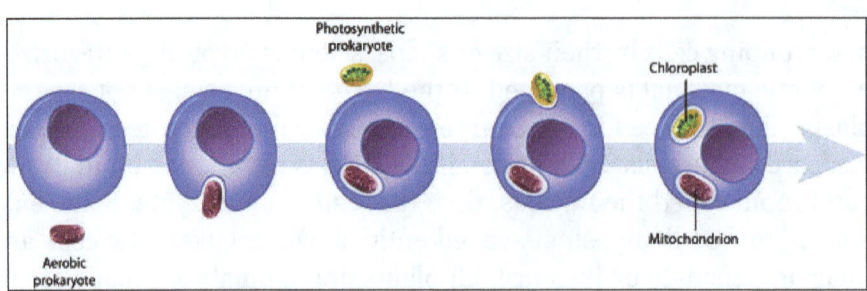

The origin of mitochondria and chloroplasts.

Mitochondria and chloroplasts likely evolved from engulfed prokaryotes that once lived as independent organisms. At some point, a eukaryotic cell engulfed an aerobic prokaryote, which then formed an endosymbiotic relationship with the host eukaryote, gradually developing into a mitochondrion. Eukaryotic cells containing mitochondria then engulfed photosynthetic prokaryotes, which evolved to become specialized chloroplast organelles.

Of course, prokaryotic cells have continued to evolve as well. Different species of bacteria and archaea have adapted to specific environments, and these prokaryotes not only survive but thrive without having their genetic material in its own compartment. For example, certain bacterial species that live in thermal vents along the ocean floor can withstand higher temperatures than any other organisms on Earth.

CELL CYCLE

Cell Cycle is the ordered sequence of events that occur in a cell in preparation for cell division. The cell cycle is a four-stage process in which the cell increases in size (gap 1, or G_1, stage), copies its DNA (synthesis, or S, stage), prepares to divide (gap 2, or G_2, stage), and divides (mitosis, or M, stage). The stages G_1, S, and G_2 make up interphase, which accounts for the span between cell divisions. On the basis of the stimulatory and inhibitory messages a cell receives, it "decides" whether it should enter the cell cycle and divide.

The proteins that play a role in stimulating cell division can be classified into four groups—growth factors, growth factor receptors, signal transducers, and nuclear regulatory proteins (transcription factors). For a stimulatory signal to reach the nucleus and "turn on" cell division, four main steps must occur. First, a growth factor must bind to its receptor on the cell membrane. Second, the receptor must become temporarily activated by this binding event. Third, this activation must stimulate a signal to be transmitted, or transduced, from the receptor at the cell surface to the nucleus within the cell. Finally, transcription factors within the nucleus must initiate the transcription of genes involved in cell proliferation. (Transcription is the process by which DNA is converted into RNA. Proteins are then made according to the RNA blueprint, and therefore transcription is crucial as an initial step in protein production).

Cells use special proteins and checkpoint signaling systems to ensure that the cell cycle progresses properly. Checkpoints at the end of G_1 and at the beginning of G_2 are designed to assess DNA for damage before and after S phase. Likewise, a checkpoint during mitosis ensures that the cell's spindle fibres are properly aligned in metaphase before the chromosomes are separated in anaphase. If DNA damage or abnormalities in spindle formation are detected at these checkpoints, the cell is forced to undergo programmed cell death, or apoptosis. However, the cell cycle and its checkpoint systems can be sabotaged by defective proteins or genes that cause malignant transformation of the cell, which can lead to cancer. For example, mutations in a protein called p53, which normally detects abnormalities in DNA at the G_1 checkpoint, can enable cancer-causing mutations to bypass this checkpoint and allow the cell to escape apoptosis.

G_0 PHASE

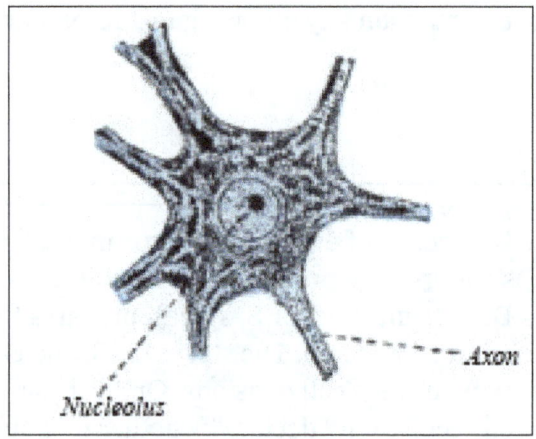

Many mammal cells, such as this 9x H neuron, remain permanently or semipermanently in G_0.

The G_0 phase describes a cellular state outside of the replicative cell cycle. Classically, cells were thought to enter G_0 primarily due to environmental factors, like nutrient deprivation, that limited the resources necessary for proliferation. Thus it was thought of as a resting phase. G_0 is now known to take different forms and occur for multiple reasons. For example, most adult neuronal cells, among the most metabolically active cells in the body, are fully differentiated and reside in a terminal G_0 phase. Neurons reside in this state, not because of stochastic or limited nutrient supply, but as a part of their internal genetic programming.

G_0 was first suggested as a cell state based on early cell cycle studies. When the first studies defined the four phases of the cell cycle using radioactive labeling techniques, it was discovered that not all cells in a population proliferate at similar rates. A population's "growth fraction" – or the fraction of the population that was growing – was actively proliferating, but other cells existed in a non-proliferative state. Some of these non-proliferating cells could respond to extrinsic stimuli and proliferate by re-entering the cell cycle. Early contrasting views either considered non-proliferating cells to simply be in an extended G_1 phase or in a cell cycle phase distinct from G_1 – termed G_0. Subsequent research pointed to a restriction point (R-point) in G_1 where cells can enter G_0 before the R-point but are committed to mitosis after the R-point. These early studies provided evidence for the existence of a G_0 state to which access is restricted. These cells that do not divide further exit G_1 phase to enter an inactive stage called quiescent stage.

Diversity of G_0 States

Three G_0 states exist and can be categorized as either reversible (quiescent) or irreversible (senescent and differentiated). Each of these three states can be entered from the G_1 phase before the cell commits to the next round of the cell cycle. Quiescence refers to a reversible G_0 state where subpopulations of cells reside in a 'quiescent' state before entering the cell cycle after activation in response to extrinsic signals. Quiescent cells are often identified by low RNA content, lack of cell proliferation markers, and increased label retention indicating low cell turnover. Senescence is distinct from quiescence because senescence is an irreversible state that cells enter in response to DNA damage or degradation that would make a cell's progeny nonviable. Such DNA damage can occur from telomere shortening over many cell divisions as well as reactive oxygen species (ROS) exposure, oncogene activation, and cell-cell fusion. While senescent cells can no longer replicate, they remain able to perform many normal cellular functions. Senescence is often a biochemical alternative to the self-destruction of such a damaged cell by apoptosis. In contrast to cellular senescence, quiescence is not a reactive event but part of the core programming of several different cell types. Finally, differentiated cells are stem cells that have progressed through a differentiation program to reach a mature – terminally differentiated – state. Differentiated cells continue to stay in G_0 and perform their main functions indefinitely.

Characteristics of Quiescent Stem Cells

Transcriptomes

The transcriptomes of several types of quiescent stem cells, such as hematopoietic, muscle, and hair follicle, have been characterized through high-throughput techniques, such as microarray and RNA sequencing. Although variations exist in their individual transcriptomes, most quiescent tissue stem cells share a common pattern of gene expression that involves downregulation of cell cycle progression genes, such as cyclin A2, cyclin B1, cyclin E2, and survivin, and upregulation of genes involved in the regulation of transcription and stem cell fate, such as FOXO3 and EZH1. Downregulation of mitochondrial cytochrome C also reflects the low metabolic state of quiescent stem cells.

Epigenetic

Many quiescent stem cells, particularly adult stem cells, also share similar epigenetic patterns. For example, H3K4me3 and H3K27me3, are two major histone methylation patterns that form a bivalent domain and are located near transcription initiation sites. These epigenetic markers have been found to regulate lineage decisions in embryonic stem cells as well as control quiescence in hair follicle and muscle stem cells via chromatin modification.

Regulation of Quiescence

Cell Cycle Regulators

Functional tumor suppressor genes, particularly p53 and Rb gene, are required to maintain stem cell quiescence and prevent exhaustion of the progenitor cell pool through excessive divisions. For example, deletion of all three components of the Rb family of proteins has been shown to halt quiescence in hematopoietic stem cells. Lack of p53 has been shown to prevent differentiation of these stem cells due to the cells' inability to exit the cell cycle into the G_0 phase. In addition to p53 and Rb, cyclin dependent kinase inhibitors (CKIs), such as p21, p27, and p57, are also important for maintaining quiescence. In mouse hematopoietic stem cells, knockout of p57 and p27 leads to G_0 exit through nuclear import of cyclin D1 and subsequent phosphorylation of Rb. Finally, the Notch signaling pathway has been shown to play an important role in maintenance of quiescence.

Post-transcriptional Regulation

Post-transcriptional regulation of gene expression via miRNA synthesis has been shown to play an equally important role in the maintenance of stem cell quiescence. miRNA strands bind to the 3' untranslated region (3' UTR) of target mRNA's, preventing their translation into functional proteins. The length of the 3' UTR of a gene determines its ability to bind to miRNA strands, thereby allowing regulation of quiescence. Some examples of miRNA's in stem cells include miR-126, which controls the PI3K/AKT/mTOR pathway in hematopoietic stem cells, miR-489, which suppresses the DEK oncogene in muscle stem cells, and miR-31, which regulates Myf5 in muscle stem cells. miRNA sequestration of mRNA within ribonucleoprotein complexes allows quiescent cells to store the mRNA necessary for quick entry into the G_1 phase.

Response to Stress

Stem cells that have been quiescent for a long time often face various environmental stressors, such as oxidative stress. However, several mechanisms allow these cells to respond to such stressors. For example, the FOXO transcription factors respond to the presence of reactive oxygen species (ROS) while HIF1A and LKB1 respond to hypoxic conditions. In hematopoietic stem cells, autophagy is induced to respond to metabolic stress.

Examples of Reversible G_0 Phase

Tissue Stem Cells

Stem cells are cells with the unique ability to produce differentiated daughter cells and to preserve their stem cell identity through self-renewal. In mammals, most adult tissues contain tissue-specific stem cells that reside in the tissue and proliferate to maintain homeostasis for the lifespan of the organism. These cells can undergo immense proliferation in response to tissue damage before differentiating and engaging in regeneration. Some tissue stem cells exist in a reversible, quiescent state indefinitely until being activated by external stimuli. Many different types of tissue stem cells exist, including muscle stem cells (MuSCs), neural stem cells (NSCs), intestinal stem cells (ISCs), and many others.

Stem cell quiescence has been recently suggested to be composed of two distinct functional phases, G_0 and an 'alert' phase termed G_{Alert}. Stem cells are believed to actively and reversibly transition between these phases to respond to injury stimuli and seem to gain enhanced tissue regenerative function in G_{Alert}. Thus, transition into G_{Alert} has been proposed as an adaptive response that enables stem cells to rapidly respond to injury or stress by priming them for cell cycle entry. In muscle stem cells, mTORC1 activity has been identified to control the transition from G_0 into G_{Alert} along with signaling through the HGF receptor cMet.

Mature Hepatocytes

While a reversible quiescent state is perhaps most important for tissue stem cells to respond quickly to stimuli and maintain proper homeostasis and regeneration, reversible G_0 phases can be found in non-stem cells such as mature hepatocytes. Hepatocytes are typically quiescent in normal livers but undergo limited replication (less than 2 cell divisions) during liver regeneration after partial hepatectomy. However, in certain cases, hepatocytes can experience immense proliferation (more than 70 cell divisions) indicating that their proliferation capacity is not hampered by existing in a reversible quiescent state.

Examples of Irreversible G_0 Phase

Senescent Cells

Often associated with aging and age-related diseases in vivo, senescent cells can be found in many renewable tissues, including the stroma, vasculature, hematopoietic system, and many epithelial organs. Resulting from accumulation over many cell divisions, senescence is often seen in age-associated degenerative phenotypes. Senescent fibroblasts in models of breast epithelial cell function have been found to disrupt milk protein production due to secretion of matrix metalloproteinases. Similarly, senescent pulmonary artery smooth muscle cells caused nearby smooth muscle cells to proliferate and migrate, perhaps contributing to hypertrophy of pulmonary arteries and eventually pulmonary hypertension.

Differentiated Muscle

During skeletal myogenesis, cycling progenitor cells known as myoblasts differentiate and fuse together into non-cycling muscle cells called myocytes that remain in a terminal G_0 phase. As a result, the fibers that make up skeletal muscle (myofibers) are cells with multiple nuclei, referred to as myonuclei, since each myonucleus originated from a single myoblast. Skeletal muscle cells continue indefinitely to provide contractile force through simultaneous contractions of cellular structures called sarcomeres. Importantly, these cells are kept in a terminal G_0 phase since disruption of muscle fiber structure after myofiber formation would prevent proper transmission of force through the length of the muscle. Muscle growth can be stimulated by growth or injury and involves the recruitment of muscle stem cells – also known as satellite cells – out of a reversible quiescent state. These stem cells differentiate and fuse to generate new muscle fibers both in parallel and in series to increase force generation capacity.

Cardiac muscle is also formed through myogenesis but instead of recruiting stem cells to fuse and form new cells, heart muscle cells – known as cardiomyocytes – simply increase in size as the heart grows larger. Similarly to skeletal muscle, if cardiomyocytes had to continue dividing to add muscle tissue the contractile structures necessary for heart function would be disrupted.

Differentiated Bone

Of the four major types of bone cells, osteocytes are the most common and also exist in a terminal G_0 phase. Osteocytes arise from osteoblasts that are trapped within a self-secreted matrix. While osteocytes also have reduced synthetic activity, they still serve bone functions besides generating structure. Osteocytes work through various mechanosensory mechanisms to assist in the routine turnover over bony matrix.

Differentiated Nerve

Outside of a few neurogenic niches in the brain, most neurons are fully differentiated and reside in a terminal G_0 phase. These fully differentiated neurons form synapses where electrical signals are transmitted by axons to the dendrites of nearby neurons. In this G_0 state, neurons continue functioning until senescence or apoptosis. Numerous studies have reported accumulation of DNA damage with age, particularly oxidative damage, in the mammalian brain.

Mechanism of G_0 Entry

Role of Rim15

Rim15 was first discovered to play a critical role in initiating meiosis in diploid yeast cells. Under conditions of low glucose and nitrogen, which are key nutrients for the survival of yeast, diploid yeast cells initiate meiosis through the activation of early meiotic-specific genes (EMGs). The expression of EMGs is regulated by Ume6. Ume6 recruits the histone deacetylases, Rpd3 and Sin3, to repress EMG expression when glucose and nitrogen levels are high, and it recruits the EMG transcription factor Ime1 when glucose and nitrogen levels are low. Rim15, named for its role in the regulation of an EMG called IME2, displaces Rpd3 and Sin3, thereby allowing Ume6 to bring Ime1 to the promoters of EMGs for meiosis initiation.

In addition to playing a role in meiosis initiation, Rim15 has also been shown to be a critical effector for yeast cell entry into G_0 in the presence of stress. Signals from several different nutrient signaling pathways converge on Rim15, which activates the transcription factors, Gis1, Msn2, and Msn4. Gis1 binds to and activates promoters containing post-diauxic growth shift (PDS) elements while Msn2 and Msn4 bind to and activate promoters containing stress-response elements (STREs). Although it is not clear how Rim15 activates Gis1 and Msn2/4, there is some speculation that it may directly phosphorylate them or be involved in chromatin remodeling. Rim15 has also been found to contain a PAS domain at its N terminal, making it a newly discovered member of the PAS kinase family. The PAS domain is a regulatory unit of the Rim15 protein that may play a role in sensing oxidative stress in yeast.

Nutrient Signaling Pathways

Glucose

Yeast grows exponentially through fermentation of glucose. When glucose levels drop, yeast shift from fermentation to cellular respiration, metabolizing the fermentative products from their exponential growth phase. This shift is known as the diauxic shift after which yeast enter G_0. When glucose levels in the surroundings are high, the production of cAMP through the RAS-cAMP-PKA pathway (a cAMP-dependent pathway) is elevated, causing protein kinase A (PKA) to inhibit its downstream target Rim15 and allow cell proliferation. When glucose levels drop, cAMP production declines, lifting PKA's inhibition of Rim15 and allowing the yeast cell to enter G_0.

Nitrogen

In addition to glucose, the presence of nitrogen is crucial for yeast proliferation. Under low nitrogen conditions, Rim15 is activated to promote cell cycle arrest through inactivation of the protein kinases TORC1 and Sch9. While TORC1 and Sch9 belong to two separate pathways, namely the TOR and Fermentable Growth Medium induced pathways respectively, both protein kinases act to promote cytoplasmic retention of Rim15. Under normal conditions, Rim15 is anchored to the cytoplasmic 14-3-3 protein, Bmh2, via phosphorylation of its Thr1075. TORC1 inactivates certain phosphatases in the cytoplasm, keeping Rim15 anchored to Bmh2, while it is thought that Sch9 promotes Rim15 cytoplasmic retention through phosphorylation of another 14-3-3 binding site close to Thr1075. When extracellular nitrogen is low, TORC1 and Sch9 are inactivated, allowing dephosphorylation of Rim15 and its subsequent transport to the nucleus, where it can activate transcription factors involved in promoting cell entry into G_0. It has also been found that Rim15 promotes its own export from the nucleus through autophosphorylation.

Phosphate

Yeast cells respond to low extracellular phosphate levels by activating genes that are involved in the production and upregulation of inorganic phosphate. The PHO pathway is involved in the regulation of phosphate levels. Under normal conditions, the yeast cyclin-dependent kinase complex, Pho80-Pho85, inactivates the Pho4 transcription factor through phosphorylation. However, when phosphate levels drop, Pho81 inhibits Pho80-Pho85, allowing Pho4 to be active. When phosphate is abundant, Pho80-Pho85 also inhibits the nuclear pool of Rim 15 by promoting phosphorylation of its Thr1075 Bmh2 binding site. Thus, Pho80-Pho85 acts in concert with Sch9 and TORC1 to promote cytoplasmic retention of Rim15 under normal conditions.

Mechanism of G_0 Exit

Cyclin C/Cdk3 and Rb

The transition from G_1 to S phase is promoted by the inactivation of Rb through its progressive hyperphosphorylation by the Cyclin D/Cdk4 and Cyclin E/Cdk2 complexes in late G_1. An early observation that loss of Rb promoted cell cycle re-entry in G_0 cells suggested that Rb is also essential in regulating the G_0 to G_1 transition in quiescent cells. Further observations revealed that levels of cyclin C mRNA are highest when human cells exit G_0, suggesting that cyclin C may be involved in Rb phosphorylation to promote cell cycle re-entry of G_0 arrested cells. Immunoprecipitation kinase assays revealed that cyclin C has Rb kinase activity. Furthermore, unlike cyclins D and E, cyclin C's Rb kinase activity is highest during early G_1 and lowest during late G_1 and S phases, suggesting that it may be involved in the G_0 to G_1 transition. The use of fluorescence-activated cell sorting to identify G_0 cells, which are characterized by a high DNA to RNA ratio relative to G_1 cells, confirmed the suspicion that cyclin C promotes G_0 exit as repression of endogenous cyclin C by RNAi in mammalian cells increased the proportion of cells arrested in G_0. Further experiments involving mutation of Rb at specific phosphorylation sites showed that cyclin C phosphorylation of Rb at S807/811 is necessary for G_0 exit. It remains unclear, however, whether this phosphorylation pattern is sufficient for G_0 exit. Finally, co-immunoprecipitation assays revealed that cyclin-dependent kinase 3 (cdk3) promotes G_0 exit by forming a complex with cyclin C to phosphorylate Rb at S807/811. Interestingly, S807/811 are also targets of cyclin D/cdk4 phosphorylation during the G_1 to S transition. This might suggest a possible compensation of cdk3 activity by cdk4, especially in light of the observation that G_0 exit is only delayed, and not permanently inhibited, in cells lacking cdk3 but functional in cdk4. Despite the overlap of phosphorylation targets, it seems that cdk3 is still necessary for the most effective transition from G_0 to G_1.

Rb and G_0 Exit

Studies suggest that Rb repression of the E2F family of transcription factors regulates the G_0 to G_1 transition just as it does the G_1 to S transition. Activating E2F complexes are associated with the recruitment of histone acetyltransferases, which activate gene expression necessary for G_1 entry, while E2F4 complexes recruit histone deacetylases, which repress gene expression. Phosphorylation of Rb by Cdk complexes allows its dissociation from E2F transcription factors and the subsequent expression of genes necessary for G_0 exit. Other members of the Rb pocket protein family, such as p107 and p130, have also been found to be involved in G_0 arrest. p130 levels are elevated in G_0 and have been found to associate with E2F-4 complexes to repress transcription of E2F target genes. Meanwhile, p107 has been found to rescue the cell arrest phenotype after loss of Rb even though p107 is expressed at comparatively low levels in G_0 cells. Taken together, these findings suggest that Rb repression of E2F transcription factors promotes cell arrest while phosphorylation of Rb leads to G_0 exit via derepression of E2F target genes. In addition to its regulation of E2F, Rb has also been shown to suppress RNA polymerase I and RNA polymerase III, which are involved in rRNA synthesis. Thus, phosphorylation of Rb also allows activation of rRNA synthesis, which is crucial for protein synthesis upon entry into G_1.

INTERPHASE

Interphase is the phase of the cell cycle in which a typical cell spends most of its life. Interphase, however is the longest stage of mitosis. During interphase, the cell copies its DNA in preparation for mitosis. Interphase is the 'daily living' or metabolic phase of the cell, in which the cell obtains nutrients and metabolizes them, grows, reads its DNA, and conducts other "normal" cell functions. This phase was formerly called the resting phase. However, interphase does not describe a cell that is merely resting; rather, the cell is living and preparing for later cell division, so the name was changed. A common misconception is that interphase is the first stage of mitosis, but since mitosis is the division of the nucleus, prophase is actually the first stage.

In interphase, the cell gets itself ready for mitosis or meiosis. Somatic cells, or normal diploid cells of the body, go through mitosis in order to reproduce themselves through cell division, whereas diploid germ cells (i.e., primary spermatocytes and primary oocytes) go through meiosis in order to create haploid gametes (i.e., sperm and ova) for the purpose of sexual reproduction.

Stages of Interphase

There are three stages of cellular interphase, with each phase ending when a cellular checkpoint checks the accuracy of the stage's completion before proceeding to the next. The stages of interphase are:

- G_0 phase.

- G_1 phase.

- G_2 phase.

G_1 Phase

The G_1 phase, or Gap 1 phase, is the first of four phases of the cell cycle that takes place in eukaryotic cell division. In this part of interphase, the cell synthesizes mRNA and proteins in preparation for subsequent steps leading to mitosis. G_1 phase ends when the cell moves into the S phase of interphase.

G_1 phase together with the S phase and G_2 phase comprise the long growth period of the cell cycle called interphase that takes place before cell division in mitosis (M phase).

During G_1 phase, the cell grows in size and synthesizes mRNA and proteins (known as histones) that are required for DNA synthesis. Once the required proteins and growth are complete, the cell enters the next phase of the cell cycle, S phase. The duration of each phase, including the G_1 phase, is different in many different types of cells. In human somatic cells, the cell cycle lasts about 18 hours, and the G_1 phase takes up about 1/3 of that time. However, in Xenopus embryos, sea urchin embryos, and Drosophila embryos, the G_1 phase is barely existent and is defined as the gap, if one exists, between the end of mitosis and the S phase.

G_1 phase and the other subphases of the cell cycle may be affected by limiting growth factors such as nutrient supply, temperature, and room for growth. Sufficient nucleotides and amino acids must be present in order to synthesize mRNA and proteins. Physiological temperatures are optimal for cell growth. In humans, the normal physiological temperature is around 37 °C (98.6 °F).

G_1 phase is particularly important in the cell cycle because it determines whether a cell commits to division or to leaving the cell cycle. If a cell is signaled to remain undivided, instead of moving onto the S phase, it will leave the G_1 phase and move into a state of dormancy called the G_0 phase. Most nonproliferating vertebrate cells will enter the G_0 phase.

Regulation

Within the cell cycle, there is a stringent set of regulations known as the cell cycle control system that controls the timing and coordination of the phases to ensure a correct order of events. Biochemical triggers known as cyclin-dependent kinases (Cdks) switch on cell cycles events at the corrected time and in the correct order to prevent any mistakes. There are three checkpoints in the cell cycle: The G_1/S Checkpoint or the Start checkpoint in yeast; the G_2/M checkpoint; and the spindle checkpoint.

Biochemical Regulators

During G_1 phase, the G_1/S cyclin activity rises significantly near the end of the G_1 phase.

Complexes of cyclin that are active during other phases of the cell cycle are kept inactivated to prevent any cell-cycle events from occurring out of order. Three methods of preventing Cdk activity are found in G_1 phase: PRB binding to E2F family transcription factors downregulate expression of S phase cyclin genes; anaphase-promoting complex (APC) is activated, which targets and degrades S and M cyclins (but not G_1/S cyclins); and a high concentration of Cdk inhibitors is found during G_1 phase.

Restriction Point

The restriction point (R) in the G_1 phase is different from a checkpoint because it does not determine whether cell conditions are ideal to move on to the next phase, but it changes the course of the cell. After a vertebrate cell has been in the G_1 phase for about three hours, the cell enters a restriction point in which it is decided whether the cell will move forward with the G_1 phase or move into the dormant G_0 phase.

This point also separates two halves of the G_1 phase; the post-mitotic and pre-mitotic phases. Between the beginning of the G_1 phase (which is also after mitosis has occurred) and R, the cell is known as being in the G_1-pm subphase, or the post-mitotic phase. After R and before S, the cell is known as being in G_1-ps, or the pre S phase interval of the G_1 phase.

In order for the cell to continue through the G_1-pm, there must be a high amount of growth factors and a steady rate of protein synthesis, otherwise the cell will move into G_0 phase.

Conflicting Research

Some authors will say that the restriction point and the G_1/S checkpoint are one and the same, but more recent studies have argued that there are two different points in the G_1 phase that check the progression of the cell. The first restriction point is growth-factor dependent and determines whether the cell moves into the G_0 phase, while the second checkpoint is nutritionally-dependent and determines whether the cell moves into the S phase. Some of the confusion between researchers has been.

The G_1/S Checkpoint

The G_1/S checkpoint is the point between G_1 phase and the S phase in which the cell is cleared for progression into the S phase. Reasons the cell would not move into the S phase include insufficient cell growth, damaged DNA, or other preparations have not been completed.

At the G_1/S checkpoint, formation of the G_1/S cyclin with Cdk to form a complex commits the cell to a new division cycle. These complexes then activate S-Cdk complexes that move forward with DNA replication in the S phase. Concurrently, anaphase-promoting complex (APC) activity decreases significantly, allowing S and M cyclins to become activated. If a cell does not clear to pass through to the S phase, it enters the dormant G_o phase in which there is no cellular growth or division.

S Phase

S phase (Synthesis Phase) is the phase of the cell cycle in which DNA is replicated, occurring between G_1 phase and G_2 phase. Since accurate duplication of the genome is critical to successful cell division, the processes that occur during S-phase are tightly regulated and widely conserved.

Regulation

Entry into S-phase is controlled by the G_1 restriction point (R), which commits cells to the remainder of the cell-cycle if there is adequate nutrients and growth signaling. This transition is essentially irreversible; after passing the restriction point, the cell will progress through S-phase even if environmental conditions become unfavorable.

Accordingly, entry into S-phase is controlled by molecular pathways that facilitate a rapid, unidirectional shift in cell state. In yeast, for instance, cell growth induces accumulation of Cln3 cyclin, which complexes with the cyclin dependent kinase CDK2. The Cln3-CDK2 complex promotes transcription of S-phase genes by inactivating the transcriptional repressor Whi5. Since upregulation of S-phase genes drive further suppression of Whi5, this pathway creates a positive feedback loop that fully commits cells to S-phase gene expression.

A remarkably similar regulatory scheme exists in mammalian cells. Mitogenic signals received throughout G_1-phase cause gradual accumulation of cyclin D, which complexes with CDK4/6. Active cyclin D-CDK4/6 complex induces release of E2F transcription factor, which in turn initiates expression of S-phase genes. Several E2F target genes promote further release of E2F, creating a positive feedback loop similar to the one found in yeast.

DNA Replication

Throughout M phase and G_1 phase, cells assemble inactive pre-replication complexes (pre-RC) on replication origins distributed throughout the genome. During S-phase, the cell converts pre-RCs into active replication forks to initiate DNA replication. This process depends on the kinase activity of Cdc7 and various S-phase CDKs, both of which are upregulated upon S-phase entry.

Activation of the pre-RC is a closely regulated and highly sequential process. After Cdc7 and S-phase CDKs phosphorylate their respective substrates, a second set of replicative factors associate with the pre-RC. Stable association encourages MCM helicase to unwind a small stretch of parental DNA into two strands of ssDNA, which in turn recruits replication protein A (RPA), an ssDNA binding protein. RPA recruitment primes the replication fork for loading of replicative DNA polymerases and PCNA sliding clamps. Loading of these factors completes the active replication fork and initiates synthesis of new DNA.

Complete replication fork assembly and activation only occurs on a small subset of replication origins. All eukaryotes possess many more replication origins than strictly needed during one cycle of DNA replication. Redundant origins may increase the flexibility of DNA replication, allowing cells to control the rate of DNA synthesis and respond to replication stress.

Histone Synthesis

Since new DNA must be packaged into nucleosomes to function properly, synthesis of canonical (non-variant) histone proteins occurs alongside DNA replication. During early S-phase, the cyclin E-Cdk2 complex phosphorylates NPAT, a nuclear coactivator of histone transcription. NPAT is activated by phosphorylation and recruits the Tip60 chromatin remodeling complex to the promoters of histone genes. Tip60 activity removes inhibitory chromatin structures and drives a three to ten-fold increase in transcription rate.

In addition to increasing transcription of histone genes, S-phase entry also regulates histone production at the RNA level. Instead of polyadenylated tails, canonical histone transcripts possess a conserved 3` stem loop motif that selective binds to Stem Loop Binding Protein (SLBP). SLBP binding is required for efficient processing, export, and translation of histone mRNAs, allowing it to function as a highly sensitive biochemical "switch". During S-phase, accumulation of SLBP acts together with NPAT to drastically increase the efficiency of histone production. However, once S-phase ends, both SLBP and bound RNA are rapidly degraded. This immediately halts histone production and prevents a toxic buildup of free histones.

Nucleosome Replication

Free histones produced by the cell during S-phase are rapidly incorporated into new nucleosomes. This process is closely tied to the replication fork, occurring immediately in "front" and "behind" the replication complex. Translocation of MCM helicase along the leading strand disrupts parental nucleosome octamers, resulting in the release of H3-H4 and H2A-H2B subunits. Reassembly of nucleosomes behind the replication fork is mediated by chromatin assembly factors (CAFs) that are loosely associated with replication proteins. Though not fully understood, the reassembly does not appear to utilize the semi-conservative scheme seen in DNA replication. Labeling experiments indicate that nucleosome duplication is predominantly conservative. The paternal H3-H4 core nucleosome remains completely segregated from newly synthesized H3-H4, resulting in the formation of nucleosomes that either contain exclusively old H3-H4 or exclusively new H3-H4. "Old" and "new" histones are assigned to each daughter strand semi-randomly, resulting in equal division of regulatory modifications.

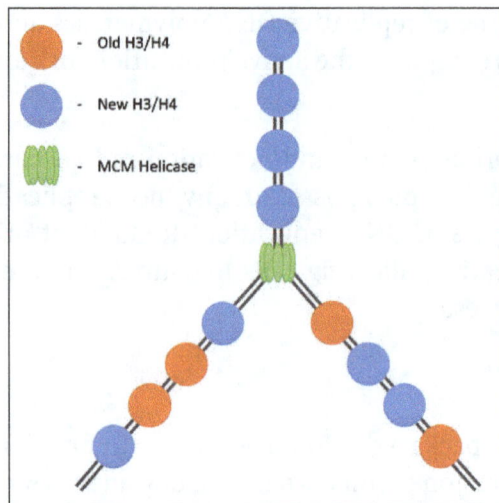

Conservative reassembly of core H3/H4 nucleosome behind the replication fork.

Reestablishment of Chromatin Domains

Immediately after division, each daughter chromatid only possesses half the epigenetic modifications present in the paternal chromatid. The cell must use this partial set of instructions to re-establish functional chromatin domains before entering mitosis.

For large genomic regions, inheritance of old H3-H4 nucleosomes is sufficient for accurate re-establishment of chromatin domains. Polycomb Repressive Complex 2 (PRC2) and several other histone-modifying complexes can "copy" modifications present on old histones onto new histones. This process amplifies epigenetic marks and counters the dilutive effect of nucleosome duplication.

However, for small domains approaching the size of individual genes, old nucleosomes are spread too thinly for accurate propagation of histone modifications. In these regions, chromatin structure is probably controlled by incorporation of histone variants during nucleosome reassembly. The close correlation seen between H3.3/H2A.Z and transcriptionally active regions lends support to this proposed mechanism. Unfortunately, a causal relationship has yet to be proven.

DNA Damage Checkpoints

During S-phase, the cell continuously scrutinizes its genome for abnormalities. Detection of DNA damage induces activation of three canonical S-phase "checkpoint pathways" that delay or arrest further cell cycle progression:

- The Replication Checkpoint detects stalled replication forks by integrating signals from RPA, ATR Interacting Protein (ATRIP), and RAD17. Upon activation, the replication checkpoint upregulates nucleotide biosynthesis and blocks replication initiation from unfired origins. Both of these processes contribute to rescue of stalled forks by increasing the availability of dNTPs.

- The S-M Checkpoint blocks mitosis until the entire genome has been successfully duplicated. This pathway induces arrest by inhibiting the Cyclin-B-CDK1 complex, which gradually accumulates throughout the cell cycle to promote mitotic entry.

- The intra-S Phase Checkpoint detects Double Strand Breaks (DSBs) through activation of ATR and ATM kinases. In addition to facilitating DNA repair, active ATR and ATM stalls cell cycle progression by promoting degradation of CDC25A, a phosphatase that removes inhibitory phosphate residues from CDKs. Homologous recombination, an accurate process for repairing DNA double-strand breaks, is most active in S phase, declines in G_2/M and is nearly absent in G_1 phase.

In addition to these canonical checkpoints, recent evidence suggests that abnormalities in histone supply and nucleosome assembly can also alter S-phase progression. Depletion of free histones in Drosophila cells dramatically prolongs S-phase and causes permanent arrest in G_2-phase. This unique arrest phenotype is not associated with activation of canonical DNA damage pathways, indicating that nucleosome assembly and histone supply may be scrutinized by a novel S-phase checkpoint.

G_2 Phase

G_2 phase, or Gap 2 phase, is the third subphase of interphase in the cell cycle directly preceding mitosis. It follows the successful completion of S phase, during which the cell's DNA is replicated. G_2 phase ends with the onset of prophase, the first phase of mitosis in which the cell's chromatin condenses into chromosomes.

G_2 phase is a period of rapid cell growth and protein synthesis during which the cell prepares itself for mitosis. Curiously, G_2 phase is not a necessary part of the cell cycle, as some cell types (particularly young Xenopus embryos and some cancers) proceed directly from DNA replication to mitosis. Though much is known about the genetic network which regulates G_2 phase and subsequent entry into mitosis, there is still much to be discovered concerning its significance and regulation, particularly in regards to cancer. One hypothesis is that the growth in G_2 phase is regulated as a method of cell size control. Fission yeast (Schizosaccharomyces pombe) has been previously shown to employ such a mechanism, via Cdr2-mediated spatial regulation of Wee1 activity. Though Wee1 is a fairly conserved negative regulator of mitotic entry, no general mechanism of cell size control in G_2 has yet been elucidated.

Biochemically, the end of G_2 phase occurs when a threshold level of active cyclin B1/CDK1 complex, also known as Maturation promoting factor (MPF) has been reached. The activity of this complex is tightly regulated during G_2. In particular, the G_2 checkpoint arrests cells in G_2 in response to DNA damage through inhibitory regulation of CDK1. It is false to say G_0 resides in G_2 phase.

Homologous Recombinational Repair

During mitotic S phase, DNA replication produces two nearly identical sister chromatids. DNA double-strand breaks that arise after replication has progressed or during the G_2 phase can be repaired before cell division occurs (M-phase of the cell cycle). Thus, during the G_2 phase, double-strand breaks in one sister chromatid may be repaired by homologous recombinational repair using the other intact sister chromatid as template.

End of G_2/Entry into Mitosis

Mitotic entry is determined by a threshold level of active cyclin-B1/CDK1 complex, also known as cyclin-B1/Cdc2 or the maturation promoting factor (MPF). Active cyclin-B1/CDK1 triggers irreversible actions in early mitosis, including centrosome separation, nuclear envelope breakdown, and spindle assembly. In vertebrates, there are five cyclin B isoforms (B1, B2, B3, B4, and B5), but the specific role of each of these isoforms in regulating mitotic entry is still unclear. It is known that cyclin B1 can compensate for loss of both cyclin B2 (and vice versa in Drosophila). Saccharomyces cerevisiae contains six B-type cyclins (Clb1-6), with Clb2 being the most essential for function. In both vertebrates and S. cerevisiae, it is speculated that the presence of multiple B-type cyclins allows different cyclins to regulate different portions of the G_2/M transition while also making the transition robust to perturbations.

Cyclin B1 Synthesis and Degradation

Cyclin B1 levels are suppressed throughout G_1 and S phases by the anaphase-promoting complex (APC), an E3 ubiquitin ligase which targets cyclin B1 for proteolysis. Transcription begins at the end of S phase after DNA replication, in response to phosphorylation of transcription factors such as NF-Y, FoxM1 and B-Myb by upstream G_1 and G_1/S cyclin-CDK complexes.

Regulation of Cyclin-B1/CDK1 Activity

Increased levels of cyclin B1 cause rising levels of cyclin B1-CDK1 complexes throughout G_2, but the complex remains inactive prior to the G_2/M transition due to inhibitory phosphorylation by the Wee1 and Myt1 kinases. Wee1is localized primarily to the nucleus and acts on the Tyr15 site, while Myt1 is localized to the outer surface of the ER and acts predominantly on the Thr14 site.

The effects of Wee1 and Myt1 are counteracted by phosphatases in the cdc25 family, which remove the inhibitory phosphates on CDK1 and thus convert the cyclin B1-CDK1 complex to its fully activated form, MPF.

Active cyclinB1-CDK1 phosphorylates and modulates the activity of Wee1 and the Cdc25 isoforms A and C. Specifically, CDK1 phosphorylation inhibits Wee1 kinase activity, activates Cdc25C phosphatase activity via activating the intermediate kinase PLK1, and stabilizes Cdc25A. Thus, CDK1 forms a positive feedback loop with Cdc25 and a double negative feedback loop with Wee1 (essentially a net positive feedback loop).

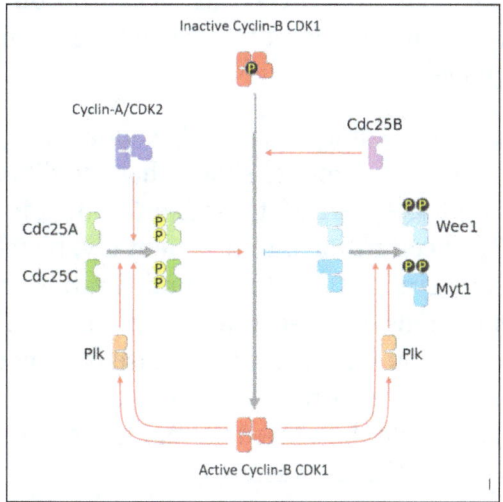

Feedback loops underlying the G_2/M transition. Cyclin-B1/CDK1 activates Plk and inactivates Wee1 and Myt1. Activated Plk activates cdc25. Activation of Cdc25 and inactivation of Wee1/Myt1 lead to further activation of Cyclin-B1/CDK1.

Positive Feedback and Switch-like Activation

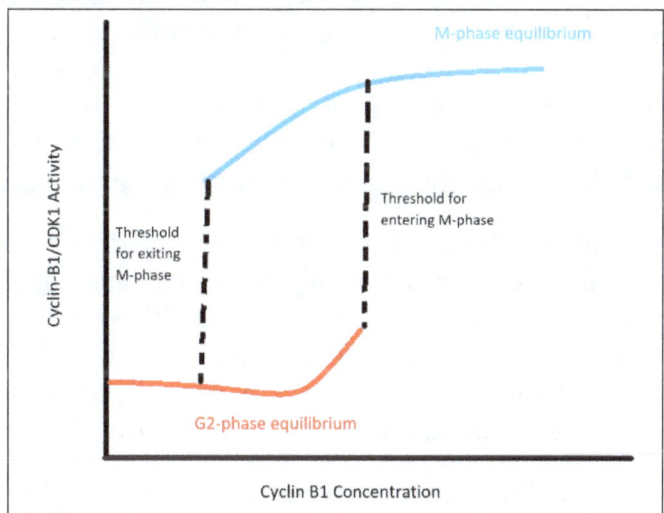

Stable equilibria for cyclin-B1/CDK1 activity at varying cyclin B1 concentrations, with the threshold of cyclin B concentration for entering mitosis higher than the threshold for exiting mitosis.

These positive feedback loops encode a hysteretic bistable switch in CDK1 activity relative to cyclin B1 levels. This switch is characterized by two distinct stable equilibria over a bistable region of cyclin B1 concentrations. One equilibrium corresponds to interphase and is characterized by inactivity of Cyclin-B1/CDK1 and Cdc25, and a high level of Wee1 and Myt1 activity. The other equilibrium corresponds to M-phase and is characterized by high activity of Cyclin-B1/CDK1 and Cdc25, and low Wee1 and Myt1 activity. Within the range of bistability, a cell's state depends upon whether it was previously in interphase or M-phase: the threshold concentration for entering M-phase is higher than the minimum concentration that will sustain M-phase activity once a cell has already exited interphase.

Scientists have both theoretically and empirically validated the bistable nature of the G_2/M transition. The Novak-Tyson model shows that the differential equations modelling the cyclin-B/CDK1-cdc25-Wee1-Myt1 feedback loop admit two stable equilibria over a range of cyclin-B concentrations. Experimentally, bistability has been validated by blocking endogenous cyclin B1 synthesis and titrating interphase and M-phase cells with varying concentrations of non-degradable cyclin B1. These experiments show that the threshold concentration for entering M-phase is higher than the threshold for exiting M-phase: nuclear envelope break-down occurs between 32-40 nm cyclin-B1 for cells exiting interphase, while the nucleus remains disintegrated at concentrations above 16-24 nm in cells already in M-phase.

This bistable, hysteretic switch is physiologically necessary for at least three reasons. First, the G_2/M transition signals the initiation of several events, such as chromosome condensation and nuclear envelope breakdown, that markedly change the morphology of the cell and are only viable in dividing cells. It is therefore essential that cyclin-B1/CDK1 activation occurs in a switch-like manner; that is, cells should rapidly settle into a discrete M-phase state after the transition, and should not persist in a continuum of intermediate states (e.g., with a partially decomposed nuclear envelope). This requirement is satisfied by the sharp discontinuity separating the interphase and M-phase equilibrium levels of CDK1 activity; as the cyclin-B concentration increases beyond the activation threshold, the cell rapidly switches to the M-phase equilibrium.

Secondly, it is also vital that the G_2/M transition occur unidirectionally, or only once per cell cycle Biological systems are inherently noisy, and small fluctuations in cyclin B1 concentrations near the threshold for the G_2/M transition should not cause the cell to switch back and forth between interphase and M-phase states. This is ensured by the bistable nature of the switch: after the cell transitions to the M-phase state, small decreases in the concentration of cyclin B do not cause the cell to switch back to interphase.

Finally, the continuation of the cell cycle requires persisting oscillations in cyclin-B/CDK1 activity as the cell and its descendants transition in and out of M-phase. Negative feedback provides one essential element of this long-term oscillation: Cyclin-B/CDK activates APC/C, which causes degradation of cyclin-B from metaphase onwards, restoring CDK1 to its inactive state. However, simple negative feedback loops lead to damped oscillations that eventually settle on a steady state. Kinetic models show that negative feedback loops coupled with bistable positive feedback motifs can lead to persistent, non-damped oscillations of the kind required for long-term cell cycling.

Positive Feedback

The positive feedback loop mentioned above, in which cyclin-B1/CDK1 promotes its own activation by inhibiting Wee1 and Myst1 and activating cdc25, does not inherently include a "trigger" mechanism to initiate the feedback loop. Recently, evidence has emerged suggesting a more important role for cyclin A2/CDK complexes in regulating the initiation of this switch. Cyclin A2/CDK2 activity begins in early S phase and increases during G_2. Cdc25B has been shown to dephosphorylate Tyr15 on CDK2 in early-to-mid G_2 in a manner similar to the aforementioned CDK1 mechanism. Downregulation of cyclin A2 in U2OS cells delays cyclin-B1/CDK1 activation by increasing Wee1 activity and lowering Plk1 and Cdc25C activity. However, cyclin A2/CDK complexes do not function strictly as activators of cyclin B1/CDK1 in G_2, as CDK2 has been shown to be required for activation of the p53-independent G_2 checkpoint activity, perhaps through a stabilizing phosphorylation on Cdc6. CDK2-/- cells also have aberrantly high levels of Cdc25A. Cyclin A2/CDK1 has also been shown to mediate proteasomal destruction of Cdc25B. These pathways are often deregulated in cancer.

Spatial Regulation

In addition to the bistable and hysteretic aspects of cyclin B1-CDK1 activation, regulation of subcellular protein localization also contributes to the G_2/M transition. Inactive cyclin B1-CDK1 accumulates in the cytoplasm, begins to be activated by cytoplasmic cdc25, and then is rapidly sequestered into the nucleus during prophase (as it is further activated). In mammals, cyclin B1/CDK1 translocation to the nucleus is activated by phosphorylation of five serine sites on cyclin B1's cytoplasmic retention site (CRS): S116, S26, S128, S133, and S147. In Xenopus laevis, cyclin B1 contains four analogous CRS serine phosphorylation sites (S94, S96, S101, and S113) indicating that this mechanism is highly conserved. Nuclear export is also inactivated by phosphorylation of cyclin B1's nuclear export signal (NES). The regulators of these phosphorylation sites are still largely unknown but several factors have been identified, including extracellular signal-regulated kinases (ERKs), PLK1, and CDK1 itself. Upon reaching some threshold level of phosphorylation, translocation of cyclin B1/CDK1 to the nucleus is extremely rapid. Once in the nucleus, cyclin B1/CDK1 phosphorylates many targets in preparation for mitosis, including histone H1, nuclear lamins, centrosomal proteins, and microtubule associated proteins (MAPs).

The subcellular localization of cdc25 also shifts from the cytosol to the nucleus during prophase. This is accomplished via removal of nuclear localization sequence (NLS)-obscuring phosphates and phosphorylation of the nuclear export signal. It is thought that the simultaneous transport of cdc25 and cyclin-B1/CDK1 into the nucleus amplify the switch-like nature of the transition by increasing the effective concentrations of the proteins.

G_2/M DNA Damage Arrest

Cells respond to DNA damage or incompletely replicated chromosomes in G_2 phase by delaying the G_2/M transition so as to prevent attempts to segregate damaged chromosomes. DNA damage is detected by the kinases ATM and ATR, which activate Chk1, an inhibitory kinase of Cdc25. Chk1 inhibits Cdc25 activity both directly and by promoting its exclusion from the nucleus. The net effect is an increase in the threshold of cyclin B1 required to initiate the hysteretic transition to M-phase, effectively stalling the cell in G_2 until the damage is repaired by mechanisms such as homology-directed repair.

Long-term maintenance of the G_2 arrest is also mediated by p53, which is stabilized in response to DNA damage. CDK1 is directly inhibited by three transcriptional targets of p53: p21, Gadd45, and 14-3-3σ. Inactive Cyclin B1/CDK1 is sequestered in the nucleus by p21, while active Cyclin B1/CDK1 complexes are sequestered in the cytoplasm by 14-3-3σ. Gadd45 disrupts the binding of Cyclin B1 and CDK1 through direct interaction with CDK1. P53 also directly transcriptionally represses CDK1.

Medical Relevance

Mutations in several genes involved in the G_2/M transition are implicated in many cancers. Overexpression of both cyclin B and CDK1, oftentimes downstream of loss of tumor suppressors such as p53, can cause an increase in cell proliferation. Experimental approaches to mitigate these changes include both pharmacological inhibition of CDK1 and downregulation of cyclin B1 expression (e.g., via siRNA).

Other attempts to modulate the G_2/M transition for chemotherapy applications have focused on the DNA damage checkpoint. Pharmacologically bypassing the G_2/M checkpoint via inhibition of Chk1 has been shown to enhance cytotoxicity of other chemotherapy drugs. Bypassing the checkpoint leads to the rapid accumulation of deleterious mutations, which is thought to drive the cancerous cells into apoptosis. Conversely, attempts to prolong the G_2/M arrest have also been shown to enhance the cytotoxicity of drugs like doxorubicin.

APOPTOSIS

Apoptosis is a form of programmed cell death that occurs in multicellular organisms. Biochemical events lead to characteristic cell changes (morphology) and death. These changes include blebbing, cell shrinkage, nuclear fragmentation, chromatin condensation, chromosomal DNA fragmentation, and global mRNA decay. The average adult human loses between 50 and 70 billion cells each day due to apoptosis. For an average human child between the ages of 8 to 14 years old approximately 20 to 30 billion cells die per day.

In contrast to necrosis, which is a form of traumatic cell death that results from acute cellular injury, apoptosis is a highly regulated and controlled process that confers advantages during an organism's life cycle. For example, the separation of fingers and toes in a developing human embryo occurs because cells between the digits undergo apoptosis. Unlike necrosis, apoptosis produces cell fragments called apoptotic bodies that phagocytic cells are able to engulf and remove before the contents of the cell can spill out onto surrounding cells and cause damage to them.

Because apoptosis cannot stop once it has begun, it is a highly regulated process. Apoptosis can be initiated through one of two pathways. In the intrinsic pathway the cell kills itself because it senses cell stress, while in the extrinsic pathway the cell kills itself because of signals from other cells. Weak external signals may also activate the intrinsic pathway of apoptosis. Both pathways induce cell death by activating caspases, which are proteases, or enzymes that degrade proteins. The two pathways both activate initiator caspases, which then activate executioner caspases, which then kill the cell by degrading proteins indiscriminately.

Activation Mechanisms

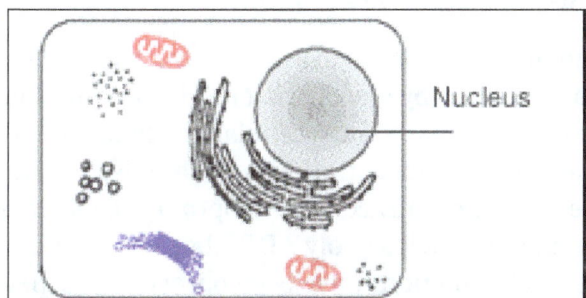
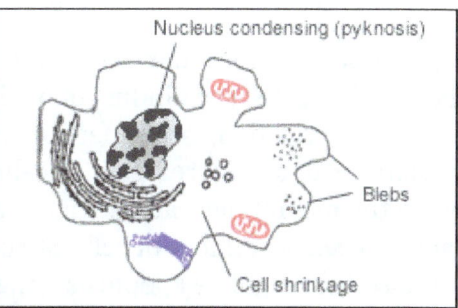
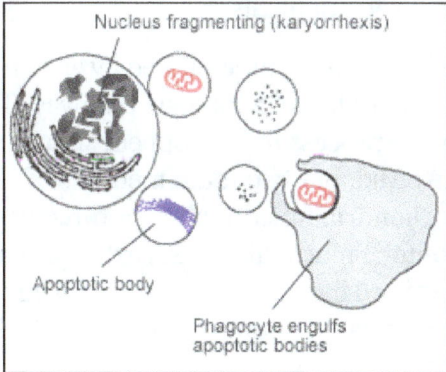

The initiation of apoptosis is tightly regulated by activation mechanisms, because once apoptosis has begun, it inevitably leads to the death of the cell. The two best-understood activation mechanisms are the intrinsic pathway (also called the mitochondrial pathway) and the extrinsic pathway. The intrinsic pathway is activated by intracellular signals generated when cells are stressed and depends on the release of proteins from the inter-membrane space of mitochondria. The extrinsic pathway is activated by extracellular ligands binding to cell-surface death receptors, which leads to the formation of the death-inducing signaling complex (DISC).

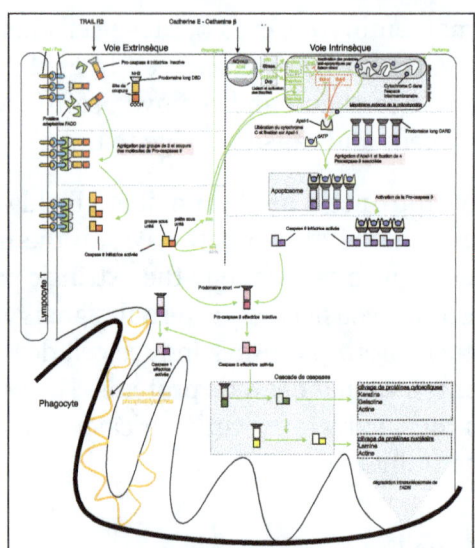

Control of the Apoptotic Mechanisms.

A cell initiates intracellular apoptotic signaling in response to a stress, which may bring about cell suicide. The binding of nuclear receptors by glucocorticoids, heat, radiation, nutrient deprivation, viral infection, hypoxia, increased intracellular concentration of free fatty acids and increased intracellular calcium concentration, for example, by damage to the membrane, can all trigger the release of intracellular apoptotic signals by a damaged cell. A number of cellular components, such as poly ADP ribose polymerase, may also help regulate apoptosis. Single cell fluctuations have been observed in experimental studies of stress induced apoptosis.

Before the actual process of cell death is precipitated by enzymes, apoptotic signals must cause regulatory proteins to initiate the apoptosis pathway. This step allows those signals to cause cell death, or the process to be stopped, should the cell no longer need to die. Several proteins are involved, but two main methods of regulation have been identified: the targeting of mitochondria functionality, or directly transducing the signal via adaptor proteins to the apoptotic mechanisms. An extrinsic pathway for initiation identified in several toxin studies is an increase in calcium concentration within a cell caused by drug activity, which also can cause apoptosis via a calcium binding protease calpain.

Intrinsic Pathway

The mitochondria are essential to multicellular life. Without them, a cell ceases to respire aerobically and quickly dies. This fact forms the basis for some apoptotic pathways. Apoptotic proteins that target mitochondria affect them in different ways. They may cause mitochondrial swelling through the formation of membrane pores, or they may increase the permeability of the mitochondrial membrane and cause apoptotic effectors to leak out. They are very closely related to intrinsic pathway, and tumors arise more frequently through intrinsic pathway than the extrinsic pathway because of sensitivity. There is also a growing body of evidence indicating that nitric oxide is able to induce apoptosis by helping to dissipate the membrane potential of mitochondria and therefore make it more permeable. Nitric oxide has been implicated in initiating and inhibiting apoptosis through its possible action as a signal molecule of subsequent pathways that activate apoptosis.

During apoptosis, cytochrome c is released from mitochondria through the actions of the proteins Bax and Bak. The mechanism of this release is enigmatic, but appears to stem from a multitude of Bax/Bak homo and hetero-dimers of Bax/Bak inserted into the outer membrane. Once cytochrome c is released it binds with Apoptotic protease activating factor – 1 (Apaf-1) and ATP, which then bind to pro-caspase-9 to create a protein complex known as an apoptosome. The apoptosome cleaves the pro-caspase to its active form of caspase-9, which in turn activates the effector caspase-3.

Mitochondria also release proteins known as SMACs (second mitochondria-derived activator of caspases) into the cell's cytosol following the increase in permeability of the mitochondria membranes. SMAC binds to proteins that inhibit apoptosis (IAPs) thereby deactivating them, and preventing the IAPs from arresting the process and therefore allowing apoptosis to proceed. IAP also normally suppresses the activity of a group of cysteine proteases called caspases, which carry out the degradation of the cell. Therefore, the actual degradation enzymes can be seen to be indirectly regulated by mitochondrial permeability.

Extrinsic Pathway

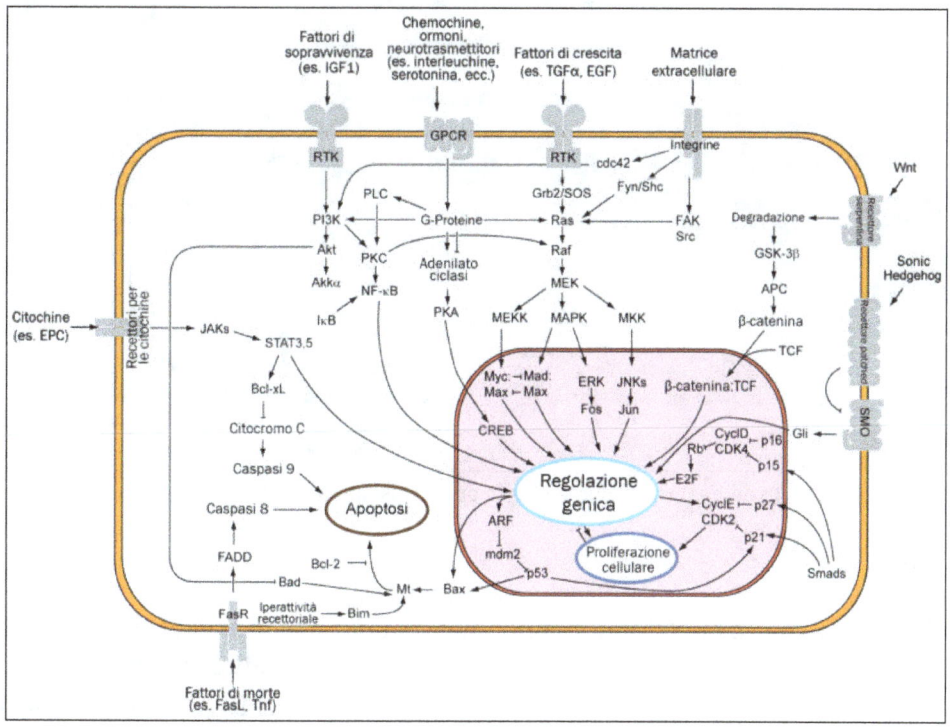

Overview of signal transduction pathways.

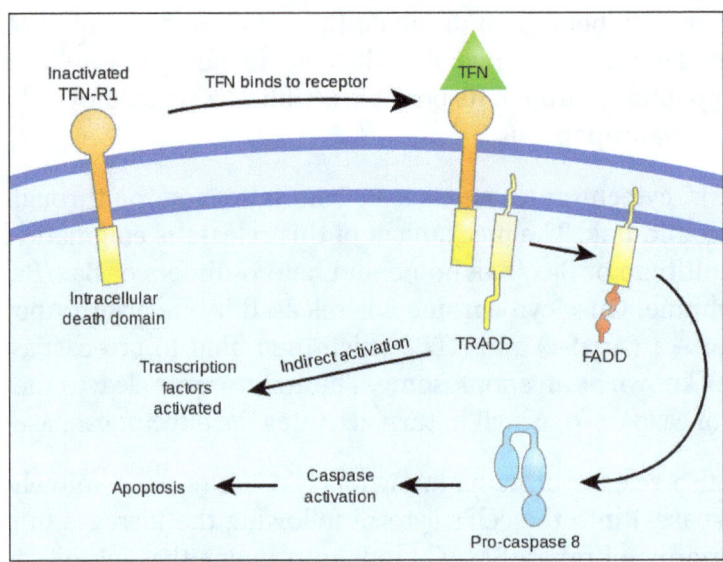

Two theories of the direct initiation of apoptotic mechanisms in mammals have been suggested: The TNF-induced (tumor necrosis factor) model and the Fas-Fas ligand-mediated model, both involving receptors of the TNF receptor (TNFR) family coupled to extrinsic signals.

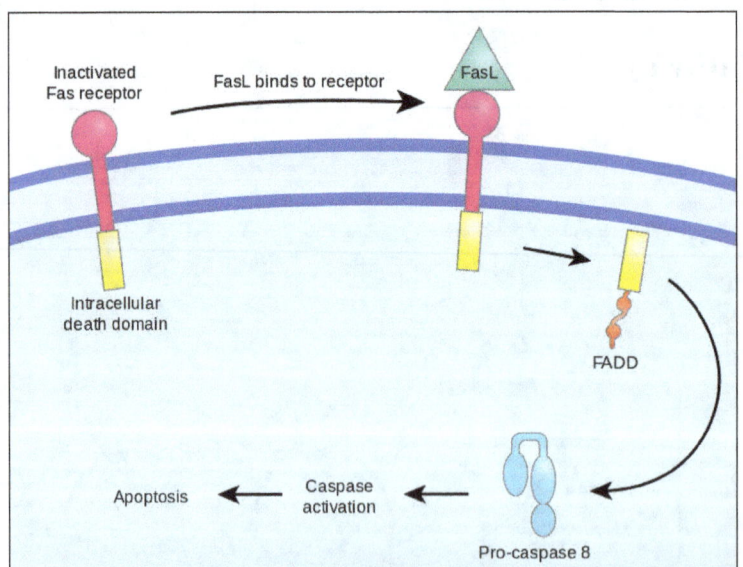

Overview of TNF (left) and Fas (right) signalling in apoptosis, an example of direct signal transduction.

TNF Path

TNF-alpha is a cytokine produced mainly by activated macrophages, and is the major extrinsic mediator of apoptosis. Most cells in the human body have two receptors for TNF-alpha: TNFR1 and TNFR2. The binding of TNF-alpha to TNFR1 has been shown to initiate the pathway that leads to caspase activation via the intermediate membrane proteins TNF receptor-associated death domain (TRADD) and Fas-associated death domain protein (FADD). cIAP1/2 can inhibit TNF-α signaling by binding to TRAF2. FLIP inhibits the activation of caspase-8. Binding of this receptor can also indirectly lead to the activation of transcription factors involved in cell survival and inflammatory responses. However, signalling through TNFR1 might also induce apoptosis in a caspase-independent manner. The link between TNF-alpha and apoptosis shows why an abnormal production of TNF-alpha plays a fundamental role in several human diseases, especially in autoimmune diseases. The TNF-alpha receptor superfamily also includes death receptors (DRs), such as DR4 and DR5. These receptors bind to the proteinTRAIL and mediate apoptosis. Apoptosis is known to be one of the primary mechanisms of targeted cancer therapy. Luminescent iridium complex-peptide hybrids (IPHs) have recently been designed, which mimic TRAIL and bind to death receptors on cancer cells, thereby inducing their apoptosis.

Fas Path

The fas receptor (First apoptosis signal) – (also known as Apo-1 or CD95) is a transmembrane protein of the TNF family which binds the Fas ligand (FasL). The interaction between Fas and FasL results in the formation of the death-inducing signaling complex (DISC), which contains the FADD, caspase-8 and caspase-10. In some types of cells (type I), processed caspase-8 directly activates other members of the caspase family, and triggers the execution of apoptosis of the cell. In other types of cells (type II), the Fas-DISC starts a feedback loop that spirals into increasing release of proapoptotic factors from mitochondria and the amplified activation of caspase-8.

Common Components

Following TNF-R1 and Fas activation in mammalian cells a balance between proapoptotic (BAX, BID, BAK, or BAD) and anti-apoptotic (Bcl-Xl and Bcl-2) members of the Bcl-2 family are established. This balance is the proportion of proapoptotic homodimers that form in the outer-membrane of the mitochondrion. The proapoptotic homodimers are required to make the mitochondrial membrane permeable for the release of caspase activators such as cytochrome c and SMAC. Control of proapoptotic proteins under normal cell conditions of nonapoptotic cells is incompletely understood, but in general, Bax or Bak are activated by the activation of BH3-only proteins, part of the Bcl-2 family.

Caspases

Caspases play the central role in the transduction of ER apoptotic signals. Caspases are proteins that are highly conserved, cysteine-dependent aspartate-specific proteases. There are two types of caspases: initiator caspases, caspase 2,8,9,10,11,12, and effector caspases, caspase 3,6,7. The activation of initiator caspases requires binding to specific oligomeric activator protein. Effector caspases are then activated by these active initiator caspases through proteolytic cleavage. The active effector caspases then proteolytically degrade a host of intracellular proteins to carry out the cell death program.

Caspase-independent Apoptotic Pathway

There also exists a caspase-independent apoptotic pathway that is mediated by AIF (apoptosis-inducing factor).

Negative Regulators of Apoptosis

Negative regulation of apoptosis inhibits cell death signaling pathways, helping tumors to evade cell death and developing drug resistance. Many families of proteins act as negative regulators categorized into either antiapoptotic factors, such as IAPs and Bcl-2 proteins or prosurvival factors like cFLIP, BNIP3, FADD, Akt, and NF-κB.

Proteolytic Caspase Cascade: Killing the Cell

Many pathways and signals lead to apoptosis, but these converge on a single mechanism that actually causes the death of the cell. After a cell receives stimulus, it undergoes organized degradation of cellular organelles by activated proteolytic caspases. In addition to the destruction of cellular organelles, mRNA is rapidly and globally degraded by a mechanism that is not yet fully characterized. mRNA decay is triggered very early in apoptosis.

A cell undergoing apoptosis shows a series of characteristic morphological changes. Early alterations include:

- Cell shrinkage and rounding occur because of the retraction lamellipodia and the breakdown of the proteinaceous cytoskeleton by caspases.

- The cytoplasm appears dense, and the organelles appear tightly packed.

- Chromatin undergoes condensation into compact patches against the nuclear envelope (also known as the perinuclear envelope) in a process known as pyknosis, a hallmark of apoptosis.

- The nuclear envelope becomes discontinuous and the DNA inside it is fragmented in a process referred to as karyorrhexis. The nucleus breaks into several discrete chromatin bodies or nucleosomal units due to the degradation of DNA.

Apoptosis progresses quickly and its products are quickly removed, making it difficult to detect or visualize on classical histology sections. During karyorrhexis, endonuclease activation leaves short DNA fragments, regularly spaced in size. These give a characteristic "laddered" appearance on agar gel after electrophoresis. Tests for DNA laddering differentiate apoptosis from ischemic or toxic cell death.

Apoptotic Cell Disassembly

Before the apoptotic cell is disposed of, there is a process of disassembly. There are three recognized steps in apoptotic cell disassembly:

- Membrane blebbing: The cell membrane shows irregular buds known as blebs. Initially these are smaller surface blebs. Later these can grow into larger so-called dynamic membrane blebs. An important regulator of apoptotic cell membrane blebbing is ROCK1 (rho associated coiled-coil-containing protein kinase 1).

- Formation of membrane protrusions: Some cell types, under specific conditions, may develop different types of long, thin extensions of the cell membrane called membrane protrusions. Three types have been described: microtubule spikes, apoptopodia (feet of death), and beaded apoptopodia (the latter having a beads-on-a-string appearance). Pannexin 1 is an important component of membrane channels involved in the formation of apoptopodia and beaded apoptopodia.

- Fragmentation: The cell breaks apart into multiple vesicles called apoptotic bodies, which undergo phagocytosis. The plasma membrane protrusions may help bring apoptotic bodies closer to phagocytes.

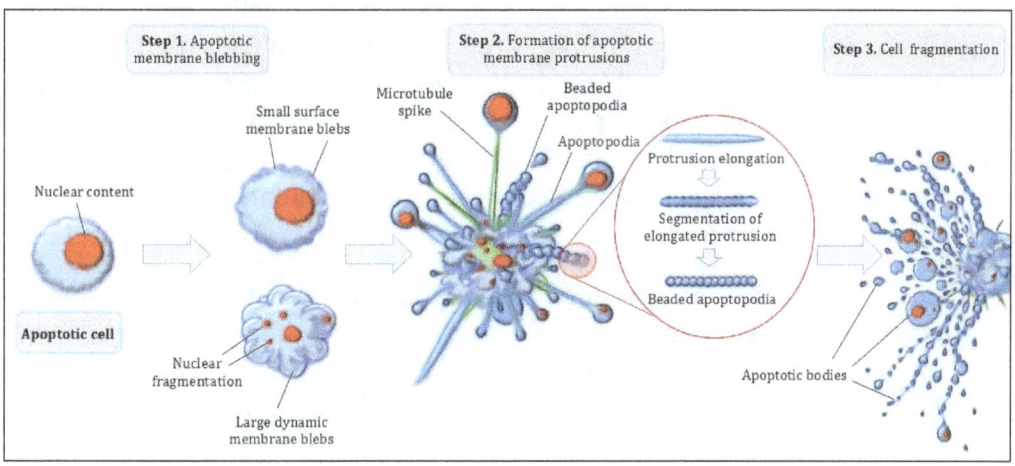

Different steps in apoptotic cell disassembly.

Removal of Dead Cells

The removal of dead cells by neighboring phagocytic cells has been termed efferocytosis. Dying cells that undergo the final stages of apoptosis display phagocytotic molecules, such as phosphatidylserine, on their cell surface. Phosphatidylserine is normally found on the inner leaflet surface of the plasma membrane, but is redistributed during apoptosis to the extracellular surface by a protein known as scramblase. These molecules mark the cell for phagocytosis by cells possessing the appropriate receptors, such as macrophages. The removal of dying cells by phagocytes occurs in an orderly manner without eliciting an inflammatory response. During apoptosis cellular RNA and DNA are separated from each other and sorted to different apoptotic bodies; separation of RNA is initiated as nucleolar segregation.

Pathway Knock-outs

Many knock-outs have been made in the apoptosis pathways to test the function of each of the proteins. Several caspases, in addition to APAF1 and FADD, have been mutated to determine the new phenotype. In order to create a tumor necrosis factor (TNF) knockout, an exon containing the nucleotides 3704–5364 was removed from the gene. This exon encodes a portion of the mature TNF domain, as well as the leader sequence, which is a highly conserved region necessary for proper intracellular processing. TNF-/- mice develop normally and have no gross structural or morphological abnormalities. However, upon immunization with SRBC (sheep red blood cells), these mice demonstrated a deficiency in the maturation of an antibody response; they were able to generate normal levels of IgM, but could not develop specific IgG levels. Apaf-1 is the protein that turns on caspase 9 by cleavage to begin the caspase cascade that leads to apoptosis. Since a -/- mutation in the APAF-1 gene is embryonic lethal, a gene trap strategy was used in order to generate an APAF-1 -/- mouse. This assay is used to disrupt gene function by creating an intragenic gene fusion. When an APAF-1 gene trap is introduced into cells, many morphological changes occur, such as spina bifida, the persistence of interdigital webs, and open brain. In addition, after embryonic day 12.5, the brain of the embryos showed several structural changes. APAF-1 cells are protected from apoptosis stimuli such as irradiation. A BAX-1 knock-out mouse exhibits normal forebrain formation and a decreased programmed cell death in some neuronal populations and in the spinal cord, leading to an increase in motor neurons.

The caspase proteins are integral parts of the apoptosis pathway, so it follows that knock-outs made have varying damaging results. A caspase 9 knock-out leads to a severe brain malformation. A caspase 8 knock-out leads to cardiac failure and thus embryonic lethality. However, with the use of cre-lox technology, a caspase 8 knock-out has been created that exhibits an increase in peripheral T cells, an impaired T cell response, and a defect in neural tube closure. These mice were found to be resistant to apoptosis mediated by CD95, TNFR, etc. but not resistant to apoptosis caused by UV irradiation, chemotherapeutic drugs, and other stimuli. Finally, a caspase 3 knock-out was characterized by ectopic cell masses in the brain and abnormal apoptotic features such as membrane blebbing or nuclear fragmentation. A remarkable feature of these KO mice is that they have a very restricted phenotype: Casp3, 9, APAF-1 KO mice have deformations of neural tissue and FADD and Casp 8 KO showed defective heart development, however in both types of KO other organs developed normally and some cell types were still sensitive to apoptotic stimuli suggesting that unknown proapoptotic pathways exist.

Methods for Distinguishing Apoptotic from Necrotic (Necroptotic) Cells

In order to perform analysis of apoptotic versus necrotic (necroptotic) cells, one can do analysis of morphology by time-lapse microscopy, flow fluorocytometry, and transmission electron microscopy. There are also various biochemical techniques for analysis of cell surface markers (phosphatidylserine exposure versus cell permeability by flow cytometry), cellular markers such as DNA fragmentation (flow cytometry), caspase activation, Bid cleavage, and cytochrome c release (Western blotting). It is important to know how primary and secondary necrotic cells can be distinguished by analysis of supernatant for caspases, HMGB1, and release of cytokeratin 18. However, no distinct surface or biochemical markers of necrotic cell death have been identified yet, and only negative markers are available. These include absence of apoptotic markers (caspase activation, cytochrome c release, and oligonucleosomal DNA fragmentation) and differential kinetics of cell death markers (phosphatidylserine exposure and cell membrane permeabilization).

Implication in Disease

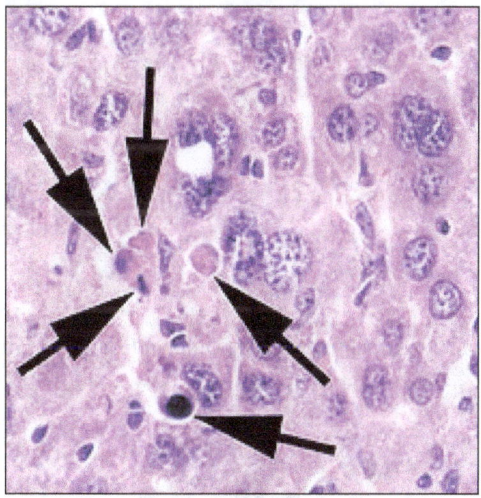

A section of mouse liver showing several apoptotic cells, indicated by arrows.

Defective Pathways

The many different types of apoptotic pathways contain a multitude of different biochemical components, many of them not yet understood. As a pathway is more or less sequential in nature, removing or modifying one component leads to an effect in another. In a living organism, this can have disastrous effects, often in the form of disease or disorder. A discussion of every disease caused by modification of the various apoptotic pathways would be impractical, but the concept overlying each one is the same: The normal functioning of the pathway has been disrupted in such a way as to impair the ability of the cell to undergo normal apoptosis. This results in a cell that lives past its "use-by-date" and is able to replicate and pass on any faulty machinery to its progeny, increasing the likelihood of the cell's becoming cancerous or diseased.

Example of this concept in action can be seen in the development of a lung cancer called NCI-H460. The X-linked inhibitor of apoptosis protein (XIAP) is overexpressed in cells of the H460 cell line. XIAPs bind to the processed form of caspase-9, and suppress the activity of apoptotic activator cytochrome c, therefore overexpression leads to a decrease in the amount of proapoptotic agonists. As a consequence, the balance of anti-apoptotic and proapoptotic effectors is upset in favour of the former, and the damaged cells continue to replicate despite being directed to die. Defects in regulation of apoptosis in cancer cells occur often at the level of control of transcription factors. As a particular example, defects in molecules that control transcription factor NF-κB in cancer change the mode of transcriptional regulation and the response to apoptotic signals, to curtail dependence on the tissue that the cell belongs. This degree of independence from external survival signals, can enable cancer metastasis.

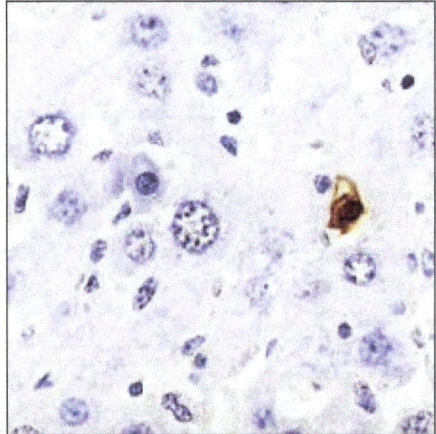

A section of mouse liver stained to show cells undergoing apoptosis (orange).

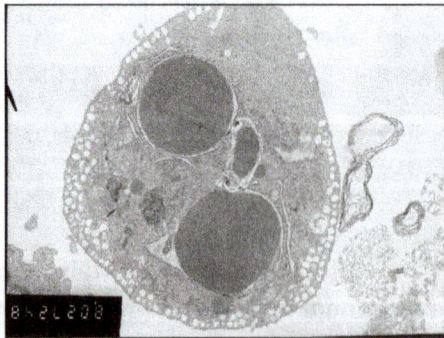

Neonatal cardiomyocytes ultrastructure after anoxia-reoxygenation.

Dysregulation of p53

The tumor-suppressor protein p53 accumulates when DNA is damaged due to a chain of biochemical factors. Part of this pathway includes alpha-interferon and beta-interferon, which induce transcription of the p53 gene, resulting in the increase of p53 protein level and enhancement of cancer cell-apoptosis. p53 prevents the cell from replicating by stopping the cell cycle at G_1, or interphase, to give the cell time to repair, however it will induce apoptosis if damage is extensive and repair efforts fail. Any disruption to the regulation of the p53 or interferon genes will result in impaired apoptosis and the possible formation of tumors.

Inhibition

Inhibition of apoptosis can result in a number of cancers, autoimmune diseases, inflammatory diseases, and viral infections. It was originally believed that the associated accumulation of cells was due to an increase in cellular proliferation, but it is now known that it is also due to a decrease in cell death. The most common of these diseases is cancer, the disease of excessive cellular proliferation, which is often characterized by an overexpression of IAP family members. As a result, the malignant cells experience an abnormal response to apoptosis induction: Cycle-regulating genes (such as p53, ras or c-myc) are mutated or inactivated in diseased cells, and further genes (such as bcl-2) also modify their expression in tumors. Some apoptotic factors are vital during mitochondrial respiration e.g. cytochrome C. Pathological inactivation of apoptosis in cancer cells is correlated with frequent respiratory metabolic shifts toward glycolysis (an observation known as the "Warburg hypothesis".

HeLa Cell

Apoptosis in HeLa cells is inhibited by proteins produced by the cell; these inhibitory proteins target retinoblastoma tumor-suppressing proteins. These tumor-suppressing proteins regulate the cell cycle, but are rendered inactive when bound to an inhibitory protein. HPV E6 and E7 are inhibitory proteins expressed by the human papillomavirus, HPV being responsible for the formation of the cervical tumor from which HeLa cells are derived. HPV E6 causes p53, which regulates the cell cycle, to become inactive. HPV E7 binds to retinoblastoma tumor suppressing proteins and limits its ability to control cell division. These two inhibitory proteins are partially responsible for HeLa cells' immortality by inhibiting apoptosis to occur. CDV (Canine Distemper Virus) is able to induce apoptosis despite the presence of these inhibitory proteins. This is an important oncolytic property of CDV: This virus is capable of killing canine lymphoma cells. Oncoproteins E6 and E7 still leave p53 inactive, but they are not able to avoid the activation of caspases induced from the stress of viral infection. These oncolytic properties provided a promising link between CDV and lymphoma apoptosis, which can lead to development of alternative treatment methods for both canine lymphoma and human non-Hodgkin lymphoma. Defects in the cell cycle are thought to be responsible for the resistance to chemotherapy or radiation by certain tumor cells, so a virus that can induce apoptosis despite defects in the cell cycle is useful for cancer treatment.

Treatments

The main method of treatment for potential death from signaling-related diseases involves either increasing or decreasing the susceptibility of apoptosis in diseased cells, depending on whether the disease is caused by either the inhibition of or excess apoptosis. For instance, treatments aim to restore apoptosis to treat diseases with deficient cell death, and to increase the apoptotic threshold to treat diseases involved with excessive cell death. To stimulate apoptosis, one can increase the number of death receptor ligands (such as TNF or TRAIL), antagonize the anti-apoptotic Bcl-2 pathway, or introduce Smac mimetics to inhibit the inhibitor (IAPs). The addition of agents such as Herceptin, Iressa, or Gleevec works to stop cells from cycling and causes apoptosis activation by blocking growth and survival signaling further upstream. Finally, adding p53-MDM2 complexes displaces p53 and activates the p53 pathway, leading to cell cycle arrest and apoptosis. Many different methods can be used either to stimulate or to inhibit apoptosis in various places along the death signaling pathway.

Apoptosis is a multi-step, multi-pathway cell-death programme that is inherent in every cell of the body. In cancer, the apoptosis cell-division ratio is altered. Cancer treatment by chemotherapy and irradiation kills target cells primarily by inducing apoptosis.

Hyperactive Apoptosis

On the other hand, loss of control of cell death (resulting in excess apoptosis) can lead to neurodegenerative diseases, hematologic diseases, and tissue damage. It is of interest to note that neurons that rely on mitochondrial respiration undergo apoptosis in neurodegenerative diseases such as Alzheimer's and Parkinson's. (an observation known as the "Inverse Warburg hypothesis"). Moreover, there is an inverse epidemiological comorbidity between neurodegenerative diseases and cancer. The progression of HIV is directly linked to excess, unregulated apoptosis. In a healthy individual, the number of CD4+ lymphocytes is in balance with the cells generated by the bone marrow; however, in HIV-positive patients, this balance is lost due to an inability of the bone marrow to regenerate CD4+ cells. In the case of HIV, CD4+ lymphocytes die at an accelerated rate through uncontrolled apoptosis, when stimulated. At the molecular level, hyperactive apoptosis can be caused by defects in signaling pathways that regulate the Bcl-2 family proteins. Increased expression of apoptotic proteins such as BIM, or their decreased proteolysis, leads to cell death, and can cause a number of pathologies, depending on the cells where excessive activity of BIM occurs. Cancer cells can escape apoptosis through mechanisms that suppress BIM expression or by increased proteolysis of BIM.

Treatments

Treatments aiming to inhibit works to block specific caspases. Finally, the Akt protein kinase promotes cell survival through two pathways. Akt phosphorylates and inhibits

Bad (a Bcl-2 family member), causing Bad to interact with the 14-3-3 scaffold, resulting in Bcl dissociation and thus cell survival. Akt also activates IKKα, which leads to NF-κB activation and cell survival. Active NF-κB induces the expression of anti-apoptotic genes such as Bcl-2, resulting in inhibition of apoptosis. NF-κB has been found to play both an antiapoptotic role and a proapoptotic role depending on the stimuli utilized and the cell type.

HIV Progression

The progression of the human immunodeficiency virus infection into AIDS is due primarily to the depletion of CD4+ T-helper lymphocytes in a manner that is too rapid for the body's bone marrow to replenish the cells, leading to a compromised immune system. One of the mechanisms by which T-helper cells are depleted is apoptosis, which results from a series of biochemical pathways:

- HIV enzymes deactivate anti-apoptotic Bcl-2. This does not directly cause cell death but primes the cell for apoptosis should the appropriate signal be received. In parallel, these enzymes activate proapoptotic procaspase-8, which does directly activate the mitochondrial events of apoptosis.

- HIV may increase the level of cellular proteins that prompt Fas-mediated apoptosis.

- HIV proteins decrease the amount of CD4 glycoprotein marker present on the cell membrane.

- Released viral particles and proteins present in extracellular fluid are able to induce apoptosis in nearby "bystander" T helper cells.

- HIV decreases the production of molecules involved in marking the cell for apoptosis, giving the virus time to replicate and continue releasing apoptotic agents and virions into the surrounding tissue.

- The infected CD4+ cell may also receive the death signal from a cytotoxic T cell.

Cells may also die as direct consequences of viral infections. HIV-1 expression induces tubular cell G_2/M arrest and apoptosis. The progression from HIV to AIDS is not immediate or even necessarily rapid; HIV's cytotoxic activity toward CD4+ lymphocytes is classified as AIDS once a given patient's CD4+ cell count falls below 200.

Viral Infection

Viral induction of apoptosis occurs when one or several cells of a living organism are infected with a virus, leading to cell death. Cell death in organisms is necessary for the normal development of cells and the cell cycle maturation. It is also important in maintaining the regular functions and activities of cells.

Viruses can trigger apoptosis of infected cells via a range of mechanisms including:

- Receptor binding.

- Activation of protein kinase R (PKR).

- Interaction with p53.

- Expression of viral proteins coupled to MHC proteins on the surface of the infected cell, allowing recognition by cells of the immune system (such as Natural Killer and cytotoxic T cells) that then induce the infected cell to undergo apoptosis.

Canine distemper virus (CDV) is known to cause apoptosis in central nervous system and lymphoid tissue of infected dogs in vivo and in vitro. Apoptosis caused by CDV is typically induced via the extrinsic pathway, which activates caspases that disrupt cellular function and eventually leads to the cells death. In normal cells, CDV activates caspase-8 first, which works as the initiator protein followed by the executioner protein caspase-3. However, apoptosis induced by CDV in HeLa cells does not involve the initiator protein caspase-8. HeLa cell apoptosis caused by CDV follows a different mechanism than that in vero cell lines. This change in the caspase cascade suggests CDV induces apoptosis via the intrinsic pathway, excluding the need for the initiator caspase-8. The executioner protein is instead activated by the internal stimuli caused by viral infection not a caspase cascade.

The Oropouche virus (OROV) is found in the family Bunyaviridae. The study of apoptosis brought on by Bunyaviridae was initiated in 1996, when it was observed that apoptosis was induced by the La Crosse virus into the kidney cells of baby hamsters and into the brains of baby mice.

OROV is a disease that is transmitted between humans by the biting midge (Culicoides paraensis). It is referred to as a zoonotic arbovirus and causes febrile illness, characterized by the onset of a sudden fever known as Oropouche fever.

The Oropouche virus also causes disruption in cultured cells – cells that are cultivated in distinct and specific conditions. An example of this can be seen in HeLa cells, whereby the cells begin to degenerate shortly after they are infected.

With the use of gel electrophoresis, it can be observed that OROV causes DNA fragmentation in HeLa cells. It can be interpreted by counting, measuring, and analyzing the cells of the Sub/ G_1 cell population. When HeLA cells are infected with OROV, the cytochrome C is released from the membrane of the mitochondria, into the cytosol of the cells. This type of interaction shows that apoptosis is activated via an intrinsic pathway.

In order for apoptosis to occur within OROV, viral uncoating, viral internalization, along with the replication of cells is necessary. Apoptosis in some viruses is activated by extracellular stimuli. However, studies have demonstrated that the OROV infection causes apoptosis to be activated through intracellular stimuli and involves the mito-chondria.

Many viruses encode proteins that can inhibit apoptosis. Several viruses encode vi-ral homologs of Bcl-2. These homologs can inhibit proapoptotic proteins such as BAX and BAK, which are essential for the activation of apoptosis. Examples of viral Bcl-2 proteins include the Epstein-Barr virus BHRF1 protein and the adenovirus E1B 19K protein. Some viruses express caspase inhibitors that inhibit caspase activity and an example is the CrmA protein of cowpox viruses. Whilst a number of viruses can block the effects of TNF and Fas. For example, the M-T2 protein of myxoma viruses can bind TNF preventing it from binding the TNF receptor and inducing a response. Further-more, many viruses express p53 inhibitors that can bind p53 and inhibit its transcrip-tional transactivation activity. As a consequence, p53 cannot induce apoptosis, since it cannot induce the expression of proapoptotic proteins. The adenovirus E1B-55K pro-tein and the hepatitis B virus HBx protein are examples of viral proteins that can per-form such a function.

Viruses can remain intact from apoptosis in particular in the latter stages of infection. They can be exported in the apoptotic bodies that pinch off from the surface of the dy-ing cell, and the fact that they are engulfed by phagocytes prevents the initiation of a host response. This favours the spread of the virus.

Caspase-independent Apoptosis

The characterization of the caspases allowed the development of caspase inhibitors, which can be used to determine whether a cellular process involves active caspases. Using these inhibitors it was discovered that cells can die while displaying a morphol-ogy similar to apoptosis without caspase activation. Later studies linked this phenom-enon to the release of AIF (apoptosis-inducing factor) from the mitochondria and its translocation into the nucleus mediated by its NLS (nuclear localization signal). Inside the mitochondria, AIF is anchored to the inner membrane. In order to be released, the protein is cleaved by a calcium-dependent calpain protease.

CELL CYCLE CHECKPOINTS

Cell cycle checkpoints are control mechanisms in eukaryotic cells which ensure proper division of the cell. Each checkpoint serves as a potential point along the cell cycle, during which the conditions of the cell are assessed, with progression through the var-ious phases of the cell cycle occurring when favorable conditions are met. Currently, there are three known checkpoints: the G_1 checkpoint, also known as the restriction or start checkpoint or (Major Checkpoint); the G_2/M checkpoint; and the metaphase checkpoint, also known as the spindle checkpoint.

All living organisms are the products of repeated rounds of cell growth and division. During this process, known as the cell cycle, a cell duplicates its contents and then divides in two. The purpose of the cell cycle is to accurately duplicate each organism's DNA and then divide the cell and its contents evenly between the two resulting cells. the cell cycle consists of four main stages: G_1, during which a cell is metabolically active and continuously grows; S phase, during which DNA replication takes place; G_2, during which cell growth continues and the cell synthesizes various proteins in preparation for division; and the M (mitosis) phase, during which the duplicated chromosomes (known as the sister chromatids) separate into two daughter nuclei, and the cell divides into two daughter cells, each with a full copy of DNA. Compared to the eukaryotic cell cycle, the prokaryotic cell cycle (known as binary fission) is relatively simple and quick: the chromosome replicates from the origin of replication, a new membrane is assembled, and the cell wall forms a septum which divides the cell into two.

The cell cycle checkpoints play an important role in the control system by sensing defects that occur during essential processes such as DNA replication or chromosome segregation, and inducing a cell cycle arrest in response until the defects are repaired. The main mechanism of action of the cell cycle checkpoints is through the regulation of the activities of a family of protein kinases known as the cyclin-dependent kinases (CDKs), which bind to different classes of regulator proteins known as cyclins, with specific cyclin-CDK complexes being formed and activated at different phases of the cell cycle. Those complexes, in turn, activate different downstream targets to promote or prevent cell cycle progression.

Restriction Point

The restriction point (R) is a point in G_1 of the animal cell cycle at which the cell becomes "committed" to the cell cycle and after which extracellular proliferation stimulants are no longer required.

Extracellular Signals

Except for early embryonic development, most cells in multicellular organisms persist in a quiescent state known as G_0, where proliferation does not occur, and cells are typically terminally differentiated; other specialized cells continue to divide into adulthood. For both of these groups of cells, a decision has been made to either exit the cell cycle and become quiescent (G_0), or to reenter G_1.

A cell's decision to enter, or reenter, the cell cycle is made before S-phase in G_1 at what is known as the restriction point, and is determined by the combination of promotional and inhibitory extracellular signals that are received and processed. Before the R-point, a cell requires these extracellular stimulants to begin progressing through the first three sub-phases of G_1 (competence, entry G_{1a}, progression G_{1b}). After the R-point has been passed in G_{1b}, however, extracellular signals are no longer required, and the cell is irreversibly committed to preparing for DNA duplication. Further progression is regulated by intracellular mechanisms. Removal of stimulants before the cell reaches the R-point may result in the cell's reversion to quiescence. Under these conditions, cells are actually set back in the cell cycle, and will require additional time (about 8 hours more than the withdrawal time in culture) after passing the restriction point to enter S phase.

Mitogen Signaling

Growth factors (e.g., PDGF, FGF, and EGF) regulate entry of cells into the cell cycle and progression to the restriction point. After passing this switch-like "point of no return," cell cycle completion is no longer dependent on the presence of mitogens. Sustained mitogen signaling promotes cell cycle entry largely through regulation of the G_1 cyclins (cyclin D1-3) and their assembly with Cdk4/6, which may be mediated in parallel through both MAPK and PI3K pathways.

MAPK Signaling Cascade

The binding of extracellular growth factors to their receptor tyrosine kinases (RTK) triggers a conformational change and promotes dimerization and autophosphorylation of tyrosine residues on the cytoplasmic tail of the RTKs. These phosphorylated tyrosine residues facilitate the docking of proteins containing an SH2-domain (e.g., Grb2), which can subsequently recruit other signaling proteins to the plasma membrane and trigger signaling kinase cascades. RTK-associated Grb2 binds Sos, which is a guanine nucleotide exchange factor that converts membrane-bound Ras to its active form (Ras-GDP Ras-GTP). Active Ras activates the MAP kinase cascade, binding and activating Raf, which phosphorylates and activates MEK, which phosphorylates and activates ERK (also known as MAPK.

Active ERK then translocates into the nucleus where it activates multiple targets, such as the transcription factor serum-response factor (SRF), resulting in expression of immediate early genes—notably the transcription factors Fos and Myc. Fos/Jun dimers comprise the transcription factor complex AP-1 and activate delayed response genes, including the major G_1 cyclin, cyclin D1. Myc also regulates expression of a wide variety of pro-proliferative and pro-growth genes, including some induction of cyclin D2 and Cdk4. Additionally, sustained ERK activity seems to be important for phosphorylation and nuclear localization of CDK2, further supporting progression through the restriction point.

PI3K Pathway Signaling

p85, another SH2-domain-containing protein, binds activated RTKs and recruits PI3K (phosphoinositide-3-kinase), phosphorylating the phospholipid PIP2 to PIP3, leading to recruitment of Akt (via its PH-domain). In addition to other pro-growth and pro-survival functions, Akt inhibits glycogen synthase kinase-3β (GSK3β), thereby preventing

GSK3β -mediated phosphorylation and subsequent degradation of cyclin D1. Akt further regulates G_1/S components by mTOR-mediated promotion of cyclin D1 translation, phosphorylation of the Cdk inhibitors p27[kip1] (preventing its nuclear import) and p21[Cip1] (decreasing stability), and inactivating phosphorylation of the transcription factor FOXO4 (which regulates p27 expression). Together, this stabilization of cyclin D1 and destabilization of Cdk inhibitors favors G_1 and G_1 /S-Cdk activity.

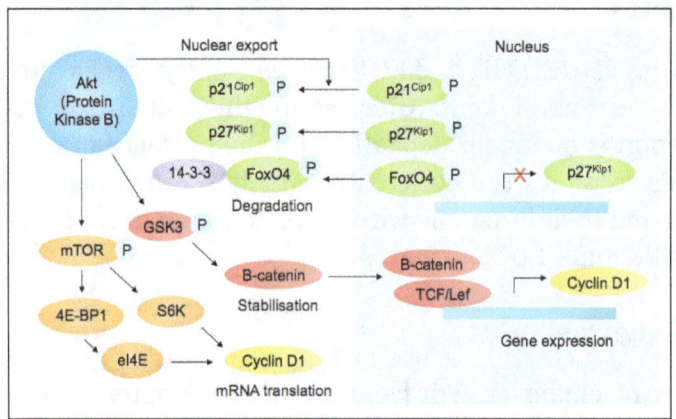

Akt signaling promotes cyclin/Cdk activity.

Anti-mitogen Signaling

Anti-mitogens like the cytokine TGF-β inhibit progression through the restriction point, causing a G_1 arrest. TGF-β signaling activates Smads, which complex with E2F4/5 to repress Myc expression and also associate with Miz1 to activate expression of the Cdk inhibitor p15[INK4b] to block cyclin D-Cdk complex formation and activity. Cells arrested with TGF-β also accumulate p21 and p27.

Mechanism

signals from extracellular growth factors are transduced in a typical manner. Growth factor binds to receptors on the cell surface, and a variety of phosphorylation cascades result in Ca^{2+} uptake and protein phosphorylation. Phosphoprotein levels are counterbalanced by phosphatases. Ultimately, transcriptional activation of certain target genes occurs. Extracellular signaling must be maintained, and the cell must also have access to sufficient nutrient supplies to support rapid protein synthesis. Accumulation of cyclin D's are essential.

Cyclin D-bound cdks 4 and 6 are activated by cdk-activating kinase and drive the cell towards the restriction point. Cyclin D, however has a high turnover rate ($t_{1/2}$ <25 min). It is because of this quick turnover rate that the cell is extremely sensitive to mitogenic signaling levels, which not only stimulate cyclin D production, but also help to stabilize cyclin D within the cell. In this way, cyclin D acts as a mitogenic signal sensor. Cdk inhibitors (CKI), such as the Ink4 proteins and p21, help to prevent improper cyclin-cdk activity.

Active cyclin D-cdk complexes phosphorylate retinoblastoma protein (pRb) in the nucleus. Unphosphorylated Rb acts as an inhibitor of G_1 by preventing E2F-mediated transcription. Once phosphorylated, E2F activates the transcription of cyclins E and A. Active cyclin E-cdk begins to accumulate and completes pRb phosphorylation, as shown in the figure.

Cdk Inhibitors and Regulation of Cyclin D/Cdk Complex Activity

p27 and p21 are stoichiometric inhibitors of G_1/S and S-cyclin-Cdk complexes. While p21 levels increase during cell-cycle entry, p27 is generally inactivated as cells progress to late G1. High cell density, mitogen starvation, and TGF-β result in accumulation of p27 and cell cycle arrest. Similarly, DNA damage and other stressors increase p21 levels, while mitogen-stimulated ERK2 and Akt activity leads to inactivating phosphorylation of p21.

Early work on p27 overexpression suggested that it can associate with and inhibit cyclin D-Cdk4/6 complexes and cyclin E/A-Cdk2 complexes in vitro and in select cell types. However, kinetic studies by LaBaer et al. found that titrating in p21 and p27 promotes assembly of the cyclin d-Cdk complex, increasing overall activity and nuclear localization of the complex. Subsequent studies elucidated that p27 may be required for cyclin D-Cdk complex formation, as p27$_{-/-}$, p21$_{-/-}$ MEFs showed a decrease in cyclin D-Cdk4 complexation that could be rescued with p27 re-expression.

Work by James et al. further suggests that phosphorylation of tyrosine residues on p27 can switch p27 between an inhibitory and non-inhibitory state while bound to cyclin D-Cdk4/6, offering a model for how p27 is capable of regulating both cyclin-Cdk complex assembly and activity. Association of p27 with cyclin D-Cdk4/6 may further promote cell cycle progression by limiting the pool of p27 available for inactivating cyclin E-Cdk2 complexes. Increasing cyclin E-Cdk2 activity in late G_1 (and cyclin A-Cdk2 in early S) leads to p21/p27 phosphorylation that promotes their nuclear export, ubiquitination, and degradation.

Spindle Checkpoint

During the process of cell division, the spindle checkpoint prevents separation of the duplicated chromosomes until each chromosome is properly attached to the spindle apparatus. In order to preserve the cell's identity and proper function, it is necessary to maintain the appropriate number of chromosomes after each cell division. An error in generating daughter cells with fewer or greater number of chromosomes than expected (a situation termed aneuploidy), may lead in best case to cell death, or alternatively it may generate catastrophic phenotypic results. Examples include:

- In cancer cells, aneuploidy is a frequent event, indicating that these cells present a defect in the machinery involved in chromosome segregation, as well as in the mechanism ensuring that segregation is correctly performed.

- In humans, Down syndrome appears in children carrying in their cells one extra copy of chromosome 21, as a result of a defect in chromosome segregation during meiosis in one of the progenitors. This defect will generate a gamete (spermatozoide or oocyte) with an extra chromosome 21. After fertilisation, this gamete will generate an embryo with three copies of chromosome 21.

The mechanisms verifying that all the requirements to pass to the next phase in the cell cycle have been fulfilled are called checkpoints. All along the cell cycle, there are different checkpoints. The checkpoint ensuring that chromosome segregation is correct is termed spindle assembly checkpoint (SAC), spindle checkpoint or mitotic checkpoint. During mitosis or meiosis, the spindle checkpoint prevents anaphase onset until all chromosomes are properly attached to the spindle. To achieve proper segregation, the two kinetochores on the sister chromatids must be attached to opposite spindle poles (bipolar orientation). Only this pattern of attachment will ensure that each daughter cell receives one copy of the chromosome.

Chromosome Segregation

Cell Division: Duplication of Material and Distribution to Daughter Cells

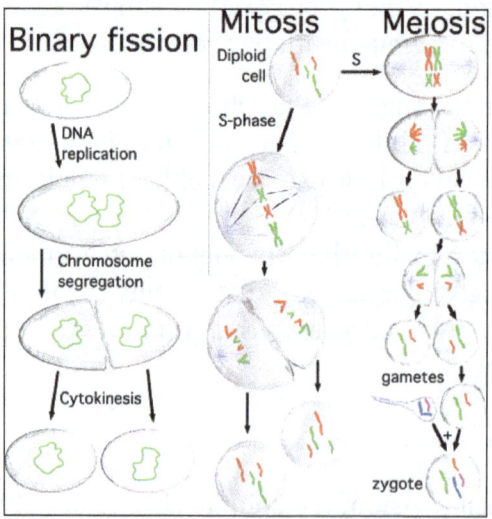

Three types of cell division: binary fission (taking place in prokaryotes), mitosis and meiosis (taking place in eukaryotes).

When cells are ready to divide, because cell size is big enough or because they receive the appropriate stimulus, they activate the mechanism to enter into the cell cycle, and they duplicate most organelles during S (synthesis) phase, including their centrosome. Therefore, when the cell division process will end, each daughter cell will receive a complete set of organelles. At the same time, during S phase all cells must duplicate their DNA very precisely, a process termed DNA replication. Once DNA replication has finished, in eukaryotes the DNA molecule is compacted and condensed, to form the mitotic chromosomes, each one constituted by two sister chromatids, which stay held together by the establishment of cohesion between them; each chromatid is a complete DNA molecule, attached via microtubules to one of the two centrosomes of the dividing cell, located at opposed poles of the cell. The structure formed by the centrosomes and the microtubules is named mitotic spindle, due to its characteristic shape, holding the chromosomes between the two centrosomes. Both sister chromatids stay together until anaphase; at this moment they separate from each other and they travel towards the centrosome to which they are attached. In this way, when the two daughter cells separate at the end of the division process, each one will receive a complete set of chromatids. The mechanism responsible for the correct distribution of sister chromatids during cell division is named chromosome segregation.

To ensure that chromosome segregation takes place correctly, cells have developed a precise and complex mechanism. In the first place, cells must coordinate centrosome duplication with DNA replication, and a failure in this coordination will generate monopolar or multipolar mitotic spindles, which generally will produce abnormal chromosome segregation, because in this case, chromosome distribution will not take place in a balanced way.

Mitosis: Anchoring of Chromosomes to the Spindle and Chromosome Segregation

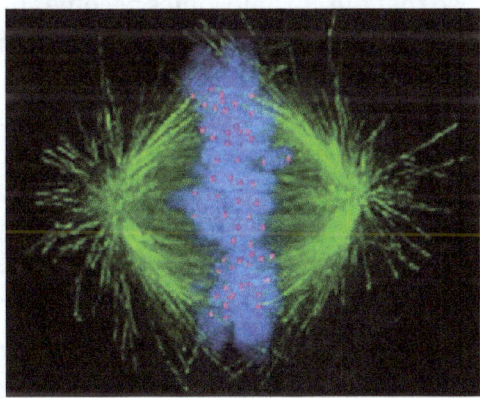

Human cell during mitosis; microtubules are shown in green (forming the mitotic spindle), chromosomes are in blue in the spindle equator and kinetochores in red.

During S phase, the centrosome starts to duplicate. Just at the beginning of mitosis, both centrioles achieve their maximal length, recruit additional material and their capacity to nucleate microtubules increases. As mitosis progresses, both centrosomes separate to generate the mitotic spindle. In this way, the mitotic spindle has two poles emanating microtubules. Microtubules (MTs) are long proteic filaments, with asymmetric extremities: one end termed "minus" (-) end, relatively stable and close to the centrosome, and an end termed "plus" (+) end, with alternating phases of growth and retraction, exploring the center of the cell searching the chromosomes. Each chromatid has a special region, named the centromere, on top of which is assembled a proteic structure termed kinetochore, which is able to stabilize the microtubule plus end. Therefore, if by chance a microtubule exploring the center of the cell encounters a kinetochore, it may happen that the kinetochore will capture it, so that the chromosome will become attached to the spindle via the kinetochore of one of its sister chromatids. The chromosome plays an active role in the attachment of kinetochores to the spindle. Bound to the chromatin is a Ran guanine nucleotide exchange factor (GEF) that stimulates cytosolic Ran near the chromosome to bind GTP in place of GDP. The activated GTP-bound form of Ran releases microtubule-stabilizing proteins, such as TPX2, from protein complexes in the cytosol, which induces nucleation and polymerization of microtubules around the chromosomes. These kinetochore-derived microtubules, along with kinesin motor proteins in the outer kinetochore, facilitate interactions with the lateral surface of a spindle pole-derived microtubule.

These lateral attachments are unstable, however, and must be converted to an end-on attachment. Conversion from lateral to end-on attachments allows the growth and shrinkage of the microtubule plus-ends to be converted into forces that push and pull chromosomes to achieve proper bi-orientation. As it happens that sister chromatids are attached together and both kinetochores are located back-to-back on both chromatids, when one kinetochore becomes attached to one centrosome, the sister kinetochore becomes exposed to the centrosome located in the opposed pole; for this reason, in most cases the second kinetochore becomes associated to the centrosome in the opposed pole, via its microtubules, so that the chromosomes become "bi-oriented", a fundamental configuration (also named amphitelic) to ensure that chromosome segregation will take place correctly when the cell will divide. Occasionally, one of the two sister kinetochores may attach simultaneously to MTs generated by both poles, a configuration named merotelic, which is not detected by the spindle checkpoint but that may generate lagging chromosomes during anaphase and, consequently, aneuploidy. Merotelic orientation (characterized by the absence of tension between sister kinetochores) is frequent at the beginning of mitosis, but the protein Aurora B (a kinase conserved from yeast to vertebrates) detects and eliminates this type of anchoring.

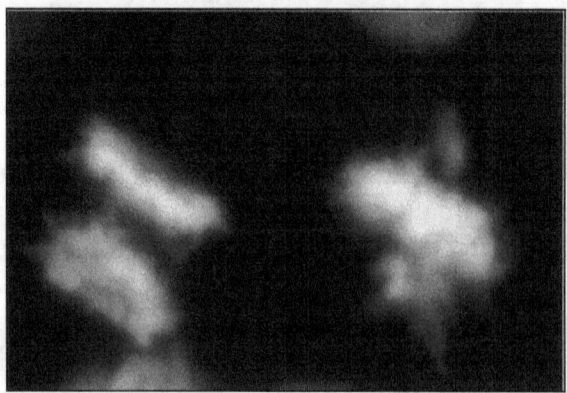

Microscopy image showing two cells with their chromosomes stained with DAPI, one at anaphase (left) and the other in metaphase (right), with most of its chromosomes in the metaphase plate and some chromosomes still not aligned.

Zirkle was one of the first researchers to observe that, when just one chromosome is retarded to arrive at the metaphase plate, anaphase onset is postponed until some minutes after its arrival. This observation, together with similar ones, suggested that a control mechanism exists at the metaphase-to-anaphase transition. Using drugs such as nocodazole and colchicine, the mitotic spindle disassembles and the cell cycle is blocked at the metaphase-to-anaphase transition. Using these drugs, the putative control mechanism was named Spindle Assembly Checkpoint (SAC). This regulatory mechanism has been intensively studied since.

Using different types of genetic studies, it has been established that diverse kinds of defects are able to activate the SAC: spindle depolymerization, the presence of dicentric chromosomes (with two centromeres), centromeres segregating in an aberrant way, defects in the spindle pole bodies in S. cerevisiae, defects in the kinetochore proteins, mutations in the centromeric DNA or defects in the molecular motors active during mitosis.

Using its own observations, Zirkle was the first to propose that "some substance, necessary for the cell to proceed to anaphase, appears some minutes after C (moment of the arrival of the last chromosome to the metaphase plate), or after a drastic change in the cytoplasmic condition, just at C or immediately after C", suggesting that this function is located on kinetochores unattached to the mitotic spindle. McIntosh extended this proposal, suggesting that one enzyme sensitive to tension located at the centromeres produces an inhibitor to the anaphase onset when the two sister kinetochores are not under bipolar tension. Indeed, the available data suggested that the signal "wait to enter in anaphase" is produced mostly on or close to unattached kinetochores. However, the primary event associated to the kinetochore attachment to the spindle, which is able to inactivate the inhibitory signal and release the metaphase arrest, could be either the acquisition of microtubules by the kinetochore, or the tension stabilizing the anchoring of microtubules to the kinetochores. Subsequent studies in cells containing two independent mitotic spindles in a sole cytoplasm showed that the inhibitor of the metaphase-to-anaphase transition is generated by unattached kinetochores and is not freely diffusible in the cytoplasm. Yet in the same study it was shown that, once the transition from metaphase to anaphase is initiated in one part of the cell, this information is extended all along the cytoplasm, and can overcome the signal "wait to enter in anaphase" associated to a second spindle containing unattached kinetochores.

Sister Chromatids Cohesion During Mitosis

Cohesin: SMC Proteins

As it has been previously noted, sister chromatids stay associated from S phase (when DNA is replicated to generate two identic copies, the two chromatids) until anaphase. At this point, the two sister chromatids separate and travel to opposite poles in the dividing cell. Genetic and biochemical studies in yeast and in egg's extracts in Xenopus laevis identified a polyprotein complex as an essential player in sister chromatids cohesion. This complex is known as the cohesin complex and in Saccharomyces cerevisiae is composed of at least four subunits: Smc1p, Smc3p, Scc1p (or Mcd1p) and Scc3p. Both Smc1p and Smc3p belong to the family of proteins for the Structural Maintenance of Chromosomes (SMC), which constitute a group of chromosomic ATPases highly conserved, and form an heterodimer (Smc1p/Smc3p). Scc1p is the homolog in S.cerevisiae of Rad21, first identified as a protein involved in DNA repair in S. pombe. These four proteins are essential in yeast, and a mutation in any of them will produce premature sister chromatid separation. In yeast, cohesin binds to preferential sites along chromosome arms, and is very abundant close to the centromeres, as it was shown in a study using chromatin immunoprecipitation.

Role of Heterochromatin

Classical cytologic observations suggested that sister chromatids are more strongly attached at heterochromatic regions, and this suggested that the special structure or composition of heterochromatin might favour cohesin recruitment. In fact, it has been shown that Swi6 (the homolog of HP-1 in S. pombe) binds to methylated Lys 9 of histone H3 and promotes the binding of cohesin to the centromeric repeats in S. pombe. More recent studies indicate that the RNAi machinery regulates heterochromatin establishment, which in turn recruits cohesin to this region, both in S. pombe and in vertebrate cells. However, there must be other mechanisms than heterochromatin to ensure an augmented cohesion at centromeres, because S. cerevisiae lacks heterochromatin next to centromeres, but the presence of a functional centromere induces an increase of cohesin association in a contiguous region, spanning 20-50kb.

In this direction, Orc2 (one protein included in the origin recognition complex, ORC, implicated in the initiation of DNA replication during S phase) is also located on kinetochores during mitosis in human cells; in agreement with this localization, some observations indicate that Orc2 in yeast is implicated in sister chromatid cohesion, and its removal induces SAC activation. It has also been observed that other components of the ORC complex (such as orc5 in S. pombe) are implicated in cohesion. However, the molecular pathway involving the ORC proteins seems to be additive to the cohesins' pathway, and it is mostly unknown.

Function of Cohesion and its Dissolution

Centromeric cohesion resists the forces exerted by spindle microtubules towards the poles, which generate tension between sister kinetochores. In turn, this tension stabilizes the attachment microtubule-kinetochore, through a mechanism implicating the protein Aurora B.

Indeed, a decrease in the cellular levels of cohesin generates the premature separation of sister chromatids, as well as defects in chromosome congression at the metaphase plate and delocalization of the proteins in the chromosomal passenger complex, which contains the protein Aurora B. The proposed structure for the cohesin complex suggests that this complex connects directly both sister chromatids. In this proposed structure, the SMC components of cohesin play a structural role, so that the SMC heterodimer may function as a DNA binding protein, whose conformation is regulated by ATP. Scc1p and Scc3p, however, would play a regulatory role.

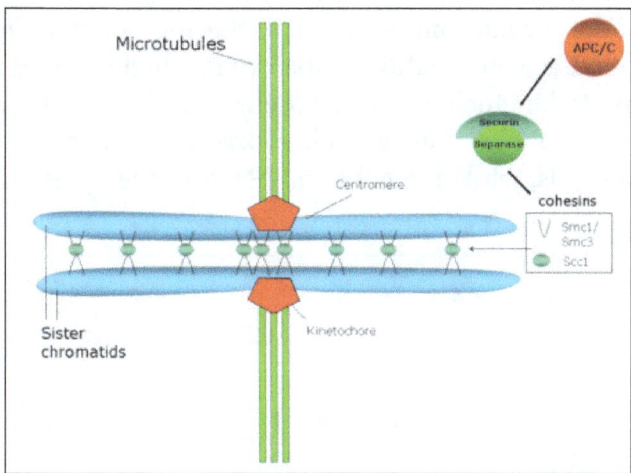

Sister chromatids cohesion, anchored to spindle microtubules via their kinetochores.

In S. cerevisiae, Pds1p (also known as securin) regulates sister chromatids cohesion, because it binds and inhibits the protease Esp1p (separin or separase). When anaphase onset is triggered, the anaphase-promoting complex (APC/C or Cyclosome) degrades securin. APC/C is a ring E3 ubiquitin ligase that recruits an E2 ubiquitin-conjugating enzyme loaded with ubiquitin. Securin is recognized only if Cdc20, the activator subunit, is bound to the APC/C core. When securin, Cdc20, and E2 are all bound to APC/C E2 ubiquitinates securin and selectively degrades it. Securin degradation releases the protease Esp1p/separase, which degrades the cohesin rings that link the two sister chromatids, therefore promoting sister chromatids separation. It has been also shown that Polo/Cdc5 kinase phosphorylates serine residues next to the cutting site for Scc1, and this phosphorylation would facilitate the cutting activity.

Although this machinery is conserved through evolution, in vertebrates most cohesin molecules are released in prophase, independently of the presence of the APC/C, in a process dependent on Polo-like 1 (PLK1) and Aurora B. Yet it has been shown that a small quantity of Scc1 remains associated to centromeres in human cells until metaphase, and a similar amount is cut in anaphase, when it disappears from centromeres. On the other hand, some experiments show that sister chromatids cohesion in the arms is lost gradually after sister centromeres have separated, and sister chromatids move toward the opposite poles of the cell.

According to some observations, a fraction of cohesins in the chromosomal arms and the centromeric cohesins are protected by the protein Shugoshin (Sgo1), avoiding their release during prophase. To be able to function as protector for the centromeric cohesion, Sg01 must be inactivated at the beginning of anaphase, as well as Pds1p. In fact, both Pds1p and Sg01 are substrates of APC/C in vertebrates.

Metaphase to Anaphase Transition

The beginning of metaphase is characterized by the connection of the microtubules to the kinetochores of the chromosomes, as well as the alignment of the chromosomes in the middle of the cell. Each chromatid has its own kinetochore, and all of the microtubules that are bound to kinetochores of sister chromatids radiate from opposite poles of the cell. These microtubules exert a pulling force on the chromosomes towards the opposite ends of the cells, while the cohesion between the sister chromatids opposes this force.

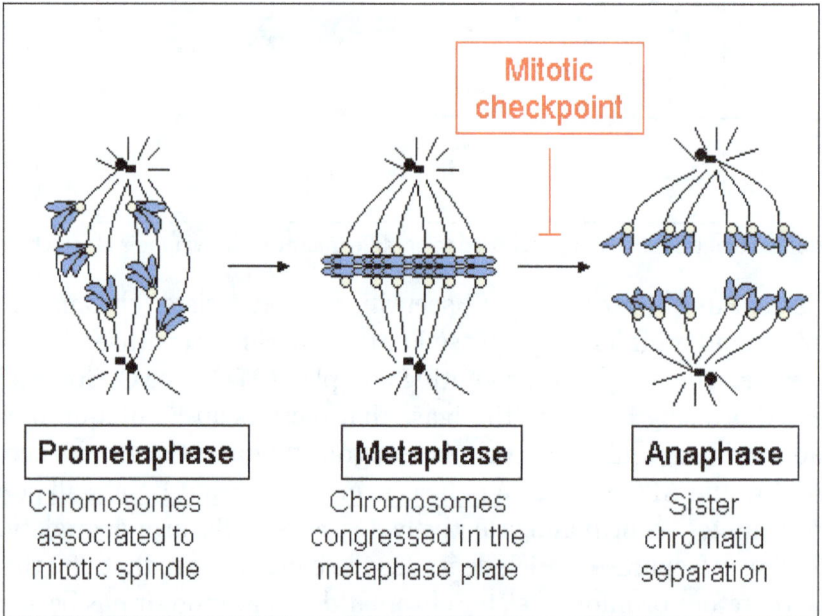

Cell cycle progression between prometaphase and anaphase.

At the metaphase to anaphase transition, this cohesion between sister chromatids is dissolved, and the separated chromatids are pulled to opposite sides of the cell by the spindle microtubules. The chromatids are further separated by the physical movement of the spindle poles themselves. Premature dissociation of the chromatids can lead to chromosome missegregation and aneuploidy in the daughter cells. Thus, the job of the metaphase checkpoint is to prevent this transition into anaphase until the chromosomes are properly attached, before the sister chromatids separate.

Spindle Assembly Checkpoint

The spindle assembly checkpoint (SAC) is an active signal produced by improperly attached kinetochores, which is conserved in all eukaryotes. The SAC stops the cell cycle by negatively regulating CDC20, thereby preventing the activation of the polyubiqui-tylation activities of anaphase promoting complex (APC). The proteins responsible for the SAC signal compose the mitotic checkpoint complex (MCC), which includes SAC proteins, MAD2/MAD3 (mitotic arrest deficient), BUB3 (budding uninhibited by ben-zimidazole), and CDC20. Other proteins involved in the SAC include MAD1, BUB1, MPS1, and Aurora B. For higher eukaryotes, additional regulators of the SAC include constituents of the ROD-ZW10 complex, p31^{comet}, MAPK, CDK1-cyclin-B, NEK2, and PLK1.

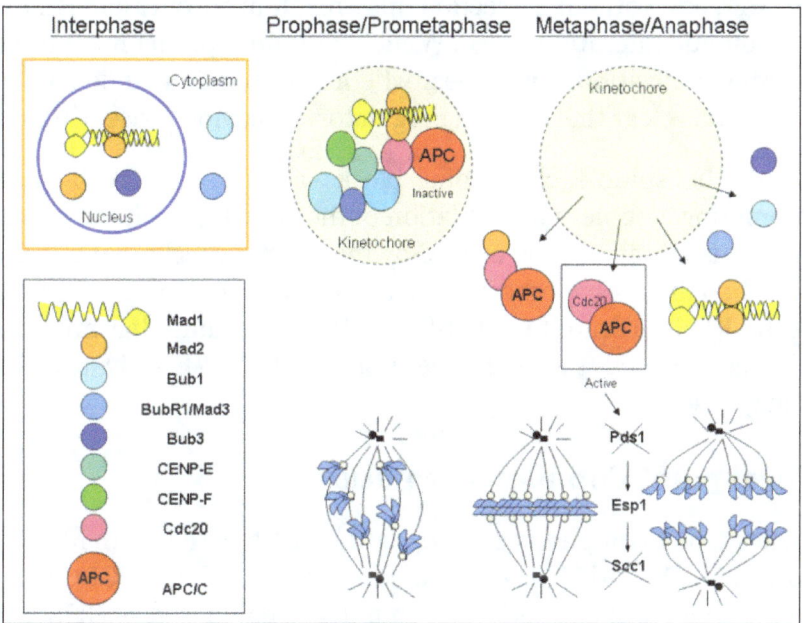

Checkpoint Activation

The SAC monitors the interaction between improperly connected kinetochores and spindle microtubules, and is maintained until kinetochores are properly attached to the spindle. During prometaphase, CDC20 and the SAC proteins concentrate at the kinetochores before attachment to the spindle assembly. These proteins keep the SAC activated until they are removed and the correct kinetochore-microtubule attachment is made. Even a single unattached kinetochore can maintain the spindle checkpoint. After attachment of microtubule plus-ends and formation of kinetochore microtubules, MAD1 and MAD2 are depleted from the kinetochore assembly. Another regulator of checkpoint activation is kinetochore tension. When sister kinetochores are properly attached to opposite spindle poles, forces in the mitotic spindle generate tension at the kinetochores. Bi-oriented sister kinetochores stabilize the kinetochore-microtubule assembly whereas weak tension has a destabilizing effect. In response to incorrect ki-netochore attachments such as syntelic attachment, where both kinetochores becomes attached to one spindle pole, the weak tension generated destabilizes the incorrect at-tachment and allows the kinetochore to reattach correctly to the spindle body. During this process, kinetochores that are attached to the mitotic spindle but that are not under tension trigger the spindle checkpoint. Aurora-B/Ipl1 kinase of the chromosomal passenger complex functions as the tensions sensor in improper kinetochore attachments. It detects and destabilizes incorrect attachments through control of the microtubule-severing KINI kinesin MCAK, the DASH complex, and the Ndc80/Hec1 complex at the microtubule-kinetochore interface. The Aurora-B/Ipl1 kinase is also critical in correcting merotelic attachments, where one kinetochore is simultaneously attached to both spindle poles. Merotelic attachments generate sufficient tension and are not detected by the SAC, and without correction, may result in chromosome mis-segregation due to slow chromatid migration speed. While microtubule attachment is independently required for SAC activation, it is unclear whether tension is an independent regulator of SAC, although it is clear that differing regulatory behaviors arise with tension.

Once activated, the spindle checkpoint blocks anaphase entry by inhibiting the anaphase-promoting complex via regulation of the activity of mitotic checkpoint complex. The mechanism of inhibition of APC by the mitotic checkpoint complex is poorly understood, although it is hypothesized that the MCC binds to APC as a pseudosubstrate using the KEN-box motif in BUBR1. At the same time that mitotic checkpoint complex is being activated, the centromere protein CENP-E activates BUBR1, which also blocks anaphase.

Mitotic Checkpoint Complex Formation

The mitotic checkpoint complex is composed of BUB3 together with MAD2 and MAD3 bound to Cdc20. MAD2 and MAD3 have distinct binding sites on CDC20, and act synergistically to inhibit APC/C. The MAD3 complex is composed of BUB3, which binds to Mad3 and BUB1B through the short linear motif known as the GLEBS motif. The exact order of attachments which must take place in order to form the MCC remains unknown. It is possible that Mad2-Cdc20 form a complex at the same time as BUBR1-BUB3-Cdc20 form another complex, and these two subcomplexes are consequently combined to form the mitotic checkpoint complex. In human cells, binding of BUBR1 to CDC20 requires prior binding of MAD2 to CDC20, so it is possible that the MAD2-CDC20 subcomplex acts as an initiator for MCC formation. BUBR1 depletion leads only to a mild reduction in Mad2-Cdc20 levels while Mad2 is required for the binding of BubR1-Bub3 to Cdc20. Nevertheless, BUBR1 is still required for checkpoint activation.

The mechanism of formation for the MCC is unclear and there are competing theories for both kinetochore-dependent and kinetochore-independent formation. In support of the kinetochore-independent theory, MCC is detectable in S. cerevisiae cells in which core kinetocore assembly proteins have been mutated and cells in which the SAC has been deactivated, which suggests that the MCC could be assembled during mitosis without kinetochore localization. In one model, unattached prometaphase kinetochores can 'sensitize' APC to inhibition of MCC by recruiting the APC to kinetochores via a functioning SAC. Furthermore, depletions of various SAC proteins have revealed that MAD2 and BUBR1 depletions affect the timing of mitosis independently of kineto-chores, while depletions of other SAC proteins result in a dysfunctional SAC without al-tering the duration of mitosis. Thus it is possible that the SAC functions through a two-stage timer where MAD2 and BUBR1 control the duration of mitosis in the first stage, which may be extended in the second stage if there are unattached kinetochores as well as other SAC proteins. However, there are lines of evidence which are in disfavor of the kinetochore-independent assembly. MCC has yet to be found during interphase, while MCC does not form from its constituents in X. laevis meiosis II extracts without the addition of sperm of nuclei and nocodazole to prevent spindle assembly.

The leading model of MCC formation is the "MAD2-template model", which depends on the kinetochore dynamics of MAD2 to create the MCC. MAD1 localizes to unattached kinetochores while binding strongly to MAD2. The localization of MAD2 and BubR1 to the kinetochore may also be dependent on the Aurora B kinase. Cells lacking Aurora B fail to arrest in metaphase even when chromosomes lack microtubule attachment. Un-attached kinetochores first bind to a MAD1-C-MAD2-p31^{comet} complex and releases the p31^{comet} through unknown mechanisms. The resulting MAD-C-MAD2 complex recruits the open conformer of Mad2 (O-Mad2) to the kinetochores. This O-Mad2 changes its conformation to closed Mad2 (C-Mad2) and binds Mad1. This Mad1/C-Mad2 complex is responsible for the recruitment of more O-Mad2 to the kinetochores, which changes its conformation to C-Mad2 and binds Cdc20 in an auto-amplification reaction. Since MAD1 and CDC20 both contain a similar MAD2-binding motif, the empty O-MAD2 conformation changes to C-MAD2 while binding to CDC20. This positive feedback loop is negatively regulated by p31^{comet}, which competitively binds to C-MAD2 bound to either MAD1 or CDC20 and reduces further O-MAD2 binding to C-MAD2. Further control mechanisms may also exist, considering that p31^{comet} is not present in lower eukaryotes. The 'template model' nomenclature is thus derived from the process where MAD1-C-MAD2 acts as a template for the formation of C-MAD2-CDC20 copies. This sequestration of Cdc20 is essential for maintaining the spindle checkpoint.

Checkpoint Deactivation

Several mechanisms exist to deactivate the SAC after correct bi-orientation of sister chromatids. Upon microtubule-kinetochore attachment, a mechanism of stripping via a dynein-dynein motor complex transports spindle checkpoint proteins away from the kinetochores. The stripped proteins, which include MAD1, MAD2, MPS1, and CENP-F, are then redistributed to the spindle poles. The stripping process is highly dependent on undamaged microtubule structure as well as dynein motility along microtubules. As well as functioning as a regulator of the C-MAD2 positive feedback loop, p31^{comet} also may act as a deactivator of the SAC. Unattached kinetochores temporarily inactivate p31^{comet}, but attachment reactivates the protein and inhibits MAD2 activation, possi-bly by inhibitory phosphorylation. Another possible mechanism of SAC inactivation results from energy-dependent dissociation of the MAD2-CDC20 complex through non-degradative ubiquitylation of CDC20. Conversely, the de-ubiquitylating enzyme protectin is required to maintain the SAC. Thus, unattached kinetochores maintain the checkpoint by continuously recreating the MAD2-CDC20 subcomplex from its components. The SAC may also be deactivated by APC activation induced proteolysis. Since the SAC is not reactivated by the loss of sister-chromatid cohesion during anaphase, the proteolysis of cyclin B and inactivation of the CDK1-cyclin-B kinase also inhibits SAC activity. Degradation of MPS1 during anaphase prevents the reactivation of SAC after removal of sister-chromatid cohesion. After checkpoint deactivation and during the normal anaphase of the cell cycle, the anaphase promoting complex is activated through decreasing MCC activity. When this happens the enzyme complex polyubiquitinates the anaphase inhibitor securin. The ubiquitination and destruction of securin at the end of metaphase releases the active protease called separase. Separase cleaves the cohesion molecules that hold the sister chromatids together to activate anaphase.

New Model for SAC Deactivation in S. Cerevisiae: The Mechanical Switch

A new mechanism has been suggested to explain how end-on microtubule attachment at the kinetochore is able to disrupt specific steps in SAC signaling. In an unattached kinetochore, the first step in the formation of the MCC is phosphorylation of Spc105 by the kinase Mps1. Phosphorylated Spc105 is then able to recruit the downstream signaling proteins Bub1 and 3; Mad 1,2, and 3; and Cdc20. Association with Mad1 at unattached kinetochores causes Mad2 to undergo a conformational change that converts it from an open form (O-Mad2) to a closed form (C-Mad2.) The C-Mad2 bound to Mad1 then dimerizes with a second O-Mad2 and catalyzes its closure around Cdc20. This C-Mad2 and Cdc20 complex, the MCC, leaves Mad1 and C-Mad2 at the kinetochore to form another MCC. The MCCs each sequester two Cdc20 molecules to prevent their interaction with the APC/C, thereby maintaining the SAC. Mps1's phosphorylation of Spc105 is both necessary and sufficient to initiate the SAC signaling pathway, but this step can only occur in the absence of microtubule attachment to the kinetochore. Endogenous Mps1 is shown to associate with the calponin-homology (CH) domain of Ndc80, which is located in the outer kinetochore region that is distant from the chromosome. Though Mps1 is docked in the outer kinetochore, it is still able to localize within the inner kinetochore and phosphorylate Spc105 because of flexible hinge regions on Ndc80. However, the mechanical switch model proposes that end-on attachment of a microtubule to the kinetochore deactivates the SAC through two mechanisms. The presence of an attached microtubule increases the distance between the Ndc80 CH domain and Spc105. Additionally, Dam1/DASH, a large complex consisting of 160 proteins that forms a ring around the attached microtubule, acts as a barrier between the two proteins. Separation prevents interactions between Mps1 and Spc105 and thus inhibits the SAC signaling pathway.

It is important to note that this model is not applicable to SAC regulation in higher order organisms, including animals. A main facet of the mechanical switch mechanism is that in S. cerevisiae the structure of the kinetochore only allows for attachment of one microtubule. Kinetochores in animals, on the other hand, are much more complex meshworks that contain binding sites for a multitude of microtubules. Microtubule attachment at all of the kinetochore binding sites is not necessary for deactivation of the SAC and progression to anaphase. Therefore, microtubule-attached and microtubule-unattached states coexist in the animal kinetochore while the SAC is inhibited. This model does not include a barrier that would prevent Mps1 associated with an attached kinetochore from phosphorylating Spc105 in an adjacent unattached kinetochore. Furthermore, the yeast Dam1/DASH complex is not present in animal cells.

Regulation

The progression of cells through the cell cycle is controlled by checkpoints at different stages. These detect if a cell contains damaged DNA and ensure those cells do not replicate. The Restriction point (R) is located at G_1 and is a key checkpoint. The vast majority of cells that pass through the R point will end up completing the entire cell cycle. Other checkpoints are located at the transitions between G_1 and S, and G_2 and M.

If damaged DNA is detected at any checkpoint, activation of the checkpoint results in increased protein p53 production. p53 is a tumour suppressor gene that stops the progression of the cell cycle and starts repair mechanisms for the damaged DNA. If this DNA cannot be repaired, then it ensures the cell undergoes apoptosis and can no longer replicate.

This cell cycle is also closely regulated by cyclins which control cell progression by activating cyclin-dependent kinase (CDK) enzymes. An example of a tumour suppressor protein would be retinoblastoma protein **(Rb)**. Rb restricts the ability of a cell to progress from G_1 to S phase in the cell cycle. CDK phosphorylates Rb to pRb, making it unable to restrict cell proliferation. This allows cells to divide normally in the cell cycle.

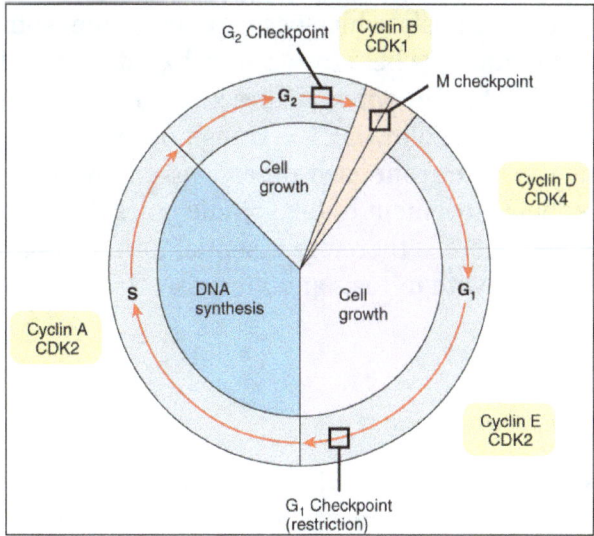

Diagram showing the cell cycle, with important checkpoints and regulators.

KINETOCHORE

A kinetochore is a disc-shaped protein structure associated with duplicated chromatids in eukaryotic cells where the spindle fibers attach during cell division to pull sister chromatids apart. The kinetochore assembles on the centromere and links the chromosome to microtubule polymers from the mitotic spindle during mitosis and meiosis. Its proteins also help to hold the sister chromatids together and play a role in chromosome editing. Details of the specific areas of origin are unknown.

Monocentric organisms, including vertebrates, fungi, and most plants, have a single centromeric region on each chromosome which assembles a single, localized kinetochore. Holocentric organisms, such as nematodes and some plants, assemble a kinetochore along the entire length of a chromosome.

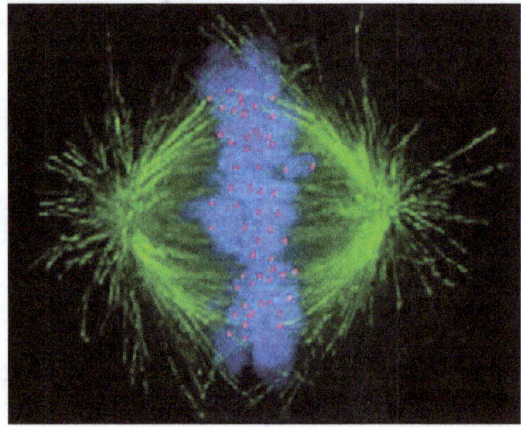

Human cell showing microtubules in green, chromosomes (DNA) in blue, and kinetochores in pink.

Kinetochores start, control, and supervise the striking movements of chromosomes during cell division. During mitosis, which occurs after chromosomes are duplicated in S phase, two sister chromatids are held together by a centromere. Each chromatid has its own kinetochore, which face in opposite directions and attach to opposite poles of the mitotic spindle apparatus. Following the transition from metaphase to anaphase, the sister chromatids separate from each other, and the individual kinetochores on each chromatid drive their movement to the spindle poles that will define the two new daughter cells. The kinetochore is therefore essential for the chromosome segregation that is classically associated with mitosis and meiosis.

Structure

The kinetochore contains two regions:

- An inner kinetochore, which is tightly associated with the centromere DNA and assembled in a specialized form of chromatin that persists throughout the cell cycle.

- An outer kinetochore, which interacts with microtubules; the outer kinetochore is a very dynamic structure with many identical components, which are assembled and functional only during cell division.

Even the simplest kinetochores consist of more than 19 different proteins. Many of these proteins are conserved between eukaryotic species, including a specialized histone H3 variant (called CENP-A or CenH3) which helps the kinetochore associate with DNA. Other proteins in the kinetochore adhere it to the microtubules (MTs) of the mitotic spindle. There are also motor proteins, including both dynein and kinesin, which generate forces that move chromosomes during mitosis. Other proteins, such as Mad2, monitor the microtubule attachment as well as the tension between sister kinetochores and activate the spindle checkpoint to arrest the cell cycle when either of these is absent.

Kinetochore functions include anchoring of chromosomes to MTs in the spindle, verification of anchoring, activation of the spindle checkpoint and participation in the generation of force to propel chromosome movement during cell division. On the other hand, microtubules are metastable polymers made of α- and β-tubulin, alternating between growing and shrinking phases, a phenomenon known as dynamic instability. MTs are highly dynamic structures, whose behavior is integrated with kinetochore function to control chromosome movement and segregation. It has also been reported that the kinetochore organization differs between mitosis and meiosis and the integrity of meiotic kinetochore is essential for meiosis specific events such as pairing of homologous chromosomes, sister kinetochore monoorientation, protection of centromeric cohesin and spindle-pole body cohesion and duplication.

In Animal Cells

The kinetochore is composed of several layers, observed initially by conventional fixation and staining methods of electron microscopy, and more recently by rapid freezing and substitution.

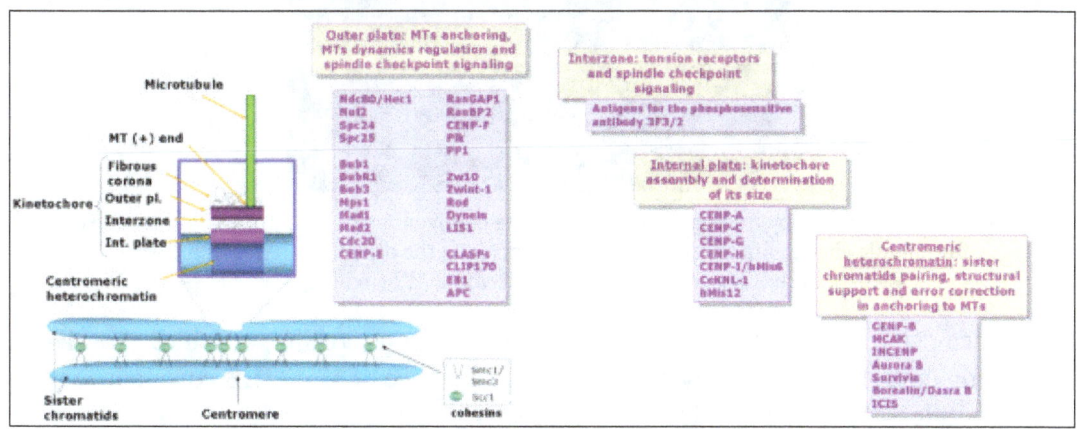

Kinetochore structure and components in vertebrate cells.

The deepest layer in the kinetochore is the inner plate, which is organized on a chromatin structure containing nucleosomes presenting a specialized histone (named CENP-A, which substitutes histone H3 in this region), auxiliary proteins, and DNA. DNA organization in the centromere (satellite DNA) is one of the least understood aspects of vertebrate kinetochores. The inner plate appears like a discrete heterochromatin domain throughout the cell cycle.

External to the inner plate is the outer plate, which is composed mostly of proteins. This structure is assembled on the surface of the chromosomes only after the nuclear envelope breaks down. The outer plate in vertebrate kinetochores contains about 20 anchoring sites for MTs (+) ends (named kMTs, after kinetochore MTs), whereas a kinetochore's outer plate in yeast (Saccharomyces cerevisiae) contains only one anchoring site.

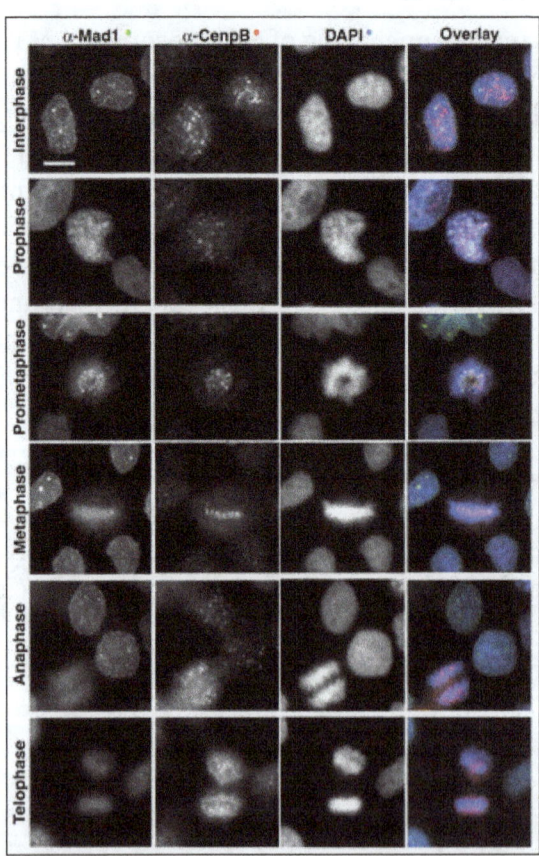

Fluorescence microscopy micrographs, showing the endogenous human protein Mad1 (one of the spindle checkpoint components) in green, along the different phases in mitosis; CENP-B, in red, is a centromeric marker, and DAPI (in blue) stains DNA.

The outermost domain in the kinetochore forms a fibrous corona, which can be visualized by conventional microscopy, yet only in the absence of MTs. This corona is formed by a dynamic network of resident and temporary proteins implicated in the spindle checkpoint, in microtubule anchoring, and in the regulation of chromosome behavior.

During mitosis, each sister chromatid forming the complete chromosome has its own kinetochore. Distinct sister kinetochores can be observed at first at the end of G_2 phase in cultured mammalian cells. These early kinetochores show a mature laminar structure before the nuclear envelope breaks down. The molecular pathway for kinetochore assembly in higher eukaryotes has been studied using gene knockouts in mice and in cultured chicken cells, as well as using RNA interference (RNAi) in C. elegans, Drosophila and human cells, yet no simple linear route can describe the data obtained so far.

The first protein to be assembled on the kinetochore is CENP-A (Cse4 in Saccharomyces cerevisiae). This protein is a specialized isoform of histone H3. CENP-A is required for incorporation of the inner kinetochore proteins CENP-C, CENP-H and CENP-I/MIS6. The relation of these proteins in the CENP-A-dependent pathway is not completely defined. For instance, CENP-C localization requires CENP-H in chicken cells, but it is independent of CENP-I/MIS6 in human cells. In C. elegans and metazoa, the incorporation of many proteins in the outer kinetochore depends ultimately on CENP-A.

Kinetochore proteins can be grouped according to their concentration at kinetochores during mitosis: Some proteins remain bound throughout cell division, whereas some others change in concentration. Furthermore, they can be recycled in their binding site on kinetochores either slowly (they are rather stable) or rapidly (dynamic):

- Proteins whose levels remain stable from prophase until late anaphase include constitutive components of the inner plate and the stable components of the outer kinetocore, such as the Ndc80 complex, KNL/KBP proteins (kinetochore-null/KNL-binding protein), MIS proteins and CENP-F. Together with the constitutive components, these proteins seem to organize the nuclear core of the inner and outer structures in the kinetochore.

- The dynamic components that vary in concentration on kinetochores during mitosis include the molecular motors CENP-E and dynein (as well as their target components ZW10 and ROD), and the spindle checkpoint proteins (such as Mad1, Mad2, BubR1 and Cdc20). These proteins assemble on the kinetochore in high concentrations in the absence of microtubules; however, the higher the number of MTs anchored to the kinetochore, the lower the concentrations of these proteins. At metaphase, CENP-E, Bub3 and Bub1 levels diminish by a factor of about three to four as compared with free kinetochores, whereas dynein/dynactin, Mad1, Mad2 and BubR1 levels are reduced by a factor of more than 10 to 100.

- Whereas the spindle checkpoint protein levels present in the outer plate diminish as MTs anchor, other components such as EB1, APC and proteins in the Ran pathway (RanGap1 and RanBP2) associate to kinetochores only when MTs are anchored. This may belong to a mechanism in the kinetochore to recognize the microtubules' plus-end (+), ensuring their proper anchoring and regulating their dynamic behavior as they remain anchored.

Function

The number of microtubules attached to one kinetochore is variable: in Saccharomyces cerevisiae only one MT binds each kinetochore, whereas in mammals there can be 15–35 MTs bound to each kinetochore. However, not all the MTs in the spindle attach to one kinetochore. There are MTs that extend from one centrosome to the other (and they are responsible for spindle length) and some shorter ones are interdigitated between the long MTs. if one breaks down the MT-kinetochore attachment using a laser beam, chromatids can no longer move, leading to an abnormal chromosome distribution. These experiments also showed that kinetochores have polarity, and that kinetochore attachment to MTs emanating from one or the other centrosome will depend on its orientation. This specificity guarantees that only one chromatid will move to each spindle side, thus ensuring the correct distribution of the genetic material. Thus, one of the basic functions of the kinetochore is the MT attachment to the spindle, which is essential to correctly segregate sister chromatids. If anchoring is incorrect, errors may ensue, generating aneuploidy, with catastrophic consequences for the cell. To prevent this from happening, there are mechanisms of error detection and correction (as the spindle assembly checkpoint), whose components reside also on the kinetochores. The movement of one chromatid towards the centrosome is produced primarily by MT depolymerization in the binding site with the kinetochore. These movements require also force generation, involving molecular motors likewise located on the kinetochores.

Chromosome Anchoring to MTs in the Mitotic Spindle

Capturing MTs

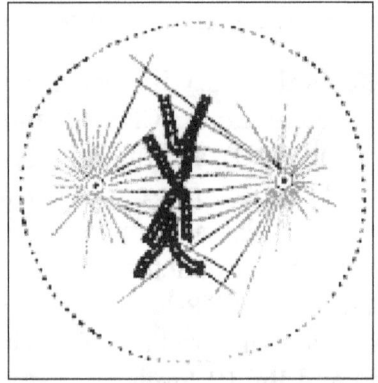

Chromosomes attach to the mitotic spindle through sister kinetochores, in a bipolar orientation.

During the synthesis phase (S phase) in the cell cycle, the centrosome starts to duplicate. Just at the beginning of mitosis, both centrioles in each centrosome reach their maximal length, centrosomes recruit additional material and their nucleation capacity for microtubules increases. As mitosis progresses, both centrosomes separate to establish the mitotic spindle. In this way, the spindle in a mitotic cell has two poles emanating microtubules. Microtubules are long proteic filaments with asymmetric extremes, a "minus"(-) end relatively stable next to the centrosome, and a "plus"(+) end enduring alternate phases of growing-shrinking, exploring the center of the cell. During this searching process, a microtubule may encounter and capture a chromosome through the kinetochore. Microtubules that find and attach a kinetochore become stabilized, whereas those microtubules remaining free are rapidly depolymerized. As chromosomes have two kinetochores associated back-to-back (one on each sister chromatid), when one of them becomes attached to the microtubules generated by one of the cellular poles, the kinetochore on the sister chromatid becomes exposed to the opposed pole; for this reason, most of the times the second kinetochore becomes attached to the microtubules emanating from the opposing pole, in such a way that chromosomes are now bi-oriented, one fundamental configuration (also termed amphitelic) to ensure the correct segregation of both chromatids when the cell will divide.

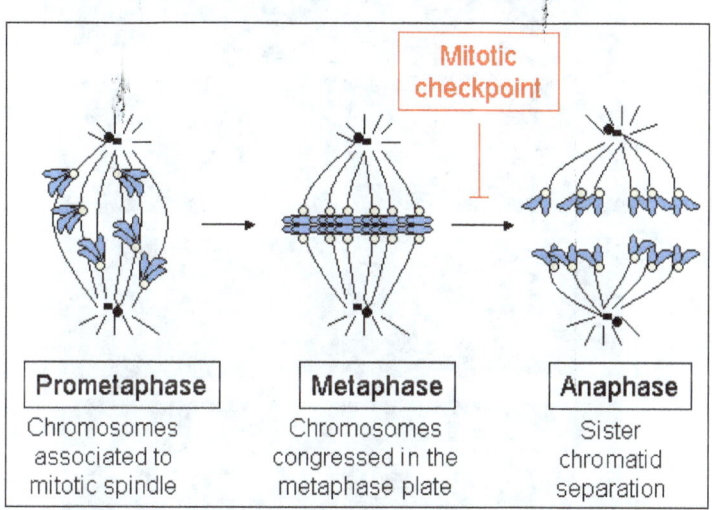

When just one microtubule is anchored to one kinetochore, it starts a rapid movement of the associated chromosome towards the pole generating that microtubule. This movement is probably mediated by the motor activity towards the "minus" (-) of the motor protein cytoplasmic dynein, which is very concentrated in the kinetochores not anchored to MTs. The movement towards the pole is slowed down as far as kinetochores acquire kMTs (MTs anchored to kinetochores) and the movement becomes directed by changes in kMTs length. Dynein is released from kinetochores as they acquire kMTs and, in cultured mammalian cells, it is required for the spindle checkpoint inactivation, but not for chromosome congression in the spindle equator, kMTs acquisition or anaphase A during chromosome segregation. In higher plants or in yeast there is no evidence of dynein, but other kinesins towards the (-) end might compensate for the lack of dynein.

Another motor protein implicated in the initial capture of MTs is CENP-E; this is a high molecular weight kinesin associated with the fibrous corona at mammalian kinetochores from prometaphase until anaphase. In cells with low levels of CENP-E, chromosomes lack this protein at their kinetochores, which quite often are defective in their ability to congress at the metaphase plate. In this case, some chromosomes may remain chronically mono-oriented (anchored to only one pole), although most chromosomes may congress correctly at the metaphase plate.

It is widely accepted that the kMTs fiber (the bundle of microtubules bound to the kinetochore) is originated by the capture of MTs polymerized at the centrosomes and spindle poles in mammalian cultured cells. However, MTs directly polymerized at kinetochores might also contribute significantly. but also to the way in which kinetochores correct defective anchoring of MTs and regulate the movement along kMTs.

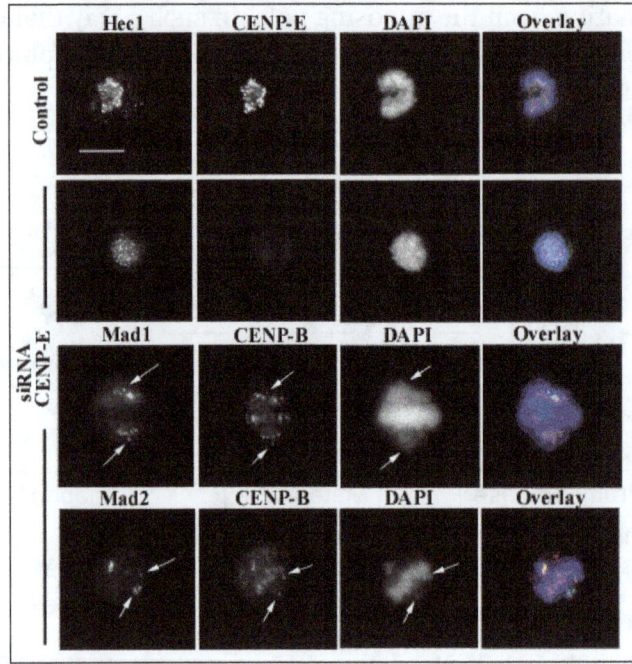

Metaphase cells with low CENP-E levels by RNAi, showing chromosomes unaligned at the metaphase plate (arrows). These chromosomes are labeled with antibodies against the mitotic checkpoint proteins Mad1/Mad2. Hec1 and CENP-B label the centromeric region (the kinetochore), and DAPI is a specific stain for DNA.

Role of Ndc80 Complex

MTs associated to kinetochores present special features: compared to free MTs, kMTs are much more resistant to cold-induced depolymerization, high hydrostatic pressures or calcium exposure. Furthermore, kMTs are recycled much more slowly than astral MTs and spindle MTs with free (+) ends, and if kMTs are released from kinetochores using a laser beam, they rapidly depolymerize.

When it was clear that neither dynein nor CENP-E is essential for kMTs formation, other molecules should be responsible for kMTs stabilization. Pioneer genetic work in yeast revealed the relevance of the Ndc80 complex in kMTs anchoring. In Saccharomyces cerevisiae, the Ndc80 complex has four components: Ndc80p, Nuf2p, Spc24p and Spc25p. Mutants lacking any of the components of this complex show loss of the kinetochore-microtubule connection, although kinetochore structure is not completely lost. Yet mutants in which kinetochore structure is lost (for instance Ndc10 mutants in yeast) are deficient both in the connection to microtubules and in the ability to activate the spindle checkpoint, probably because kinetochores work as a platform in which the components of the response are assembled.

The Ndc80 complex is highly conserved and it has been identified in S. pombe, C. elegans, Xenopus, chicken and humans. Studies on Hec1 (highly expressed in cancer cells), the human homolog of Ndc80p, show that it is important for correct chromosome congression and mitotic progression, and that it interacts with components of the cohesin and condensin complexes.

Different laboratories have shown that the Ndc80 complex is essential for stabilization of the kinetochore-microtubule anchoring, required to support the centromeric tension implicated in the establishment of the correct chromosome congression in high eukaryotes. Cells with impaired function of Ndc80 (using RNAi, gene knockout, or antibody microinjection) have abnormally long spindles, lack of tension between sister kinetochores, chromosomes unable to congregate at the metaphase plate and few or any associated kMTs.

There is a variety of strong support for the ability of the Ndc80 complex to directly associate with microtubules and form the core conserved component of the kinetochore-microtubule interface. However, formation of robust kinetochore-microtubule interactions may also require the function of additional proteins. In yeast, this connection requires the presence of the complex Dam1-DASH-DDD. Some members of this complex bind directly to MTs, whereas some others bind to the Ndc80 complex. This means that the complex Dam1-DASH-DDD might be an essential adapter between kinetochores and microtubules. However, in animals an equivalent complex has not been identified, and this question remains under intense investigation.

Verification of Kinetochore – MT Anchoring

During S-Phase, the cell duplicates all the genetic information stored in the chromosomes, in the process termed DNA replication. At the end of this process, each chromosome includes two sister chromatids, which are two complete and identical DNA molecules. Both chromatids remain associated by cohesin complexes until anaphase, when chromosome segregation occurs. If chromosome segregation happens correctly, each daughter cell receives a complete set of chromatids, and for this to happen each sister chromatid has to anchor (through the corresponding kinetochore) to MTs generated in opposed poles of the mitotic spindle. This configuration is termed amphitelic or bi-orientation.

However, during the anchoring process some incorrect configurations may also appear:

- Monotelic: Only one of the chromatids is anchored to MTs, the second kinetochore is not anchored; in this situation, there is no centromeric tension, and the spindle checkpoint is activated, delaying entry in anaphase and allowing time for the cell to correct the error. If it is not corrected, the unanchored chromatid might randomly end in any of the two daughter cells, generating aneuploidy: one daughter cell would have chromosomes in excess and the other would lack some chromosomes.

- Syntelic: Both chromatids are anchored to MTs emanating from the same pole; this situation does not generate centromeric tension either, and the spindle checkpoint will be activated. If it is not corrected, both chromatids will end in the same daughter cell, generating aneuploidy.

- Merotelic: At least one chromatid is anchored simultaneously to MTs emanating from both poles. This situation generates centromeric tension, and for this reason the spindle checkpoint is not activated. If it is not corrected, the chromatid bound to both poles will remain as a lagging chromosome at anaphase, and finally will be broken in two fragments, distributed between the daughter cells, generating aneuploidy.

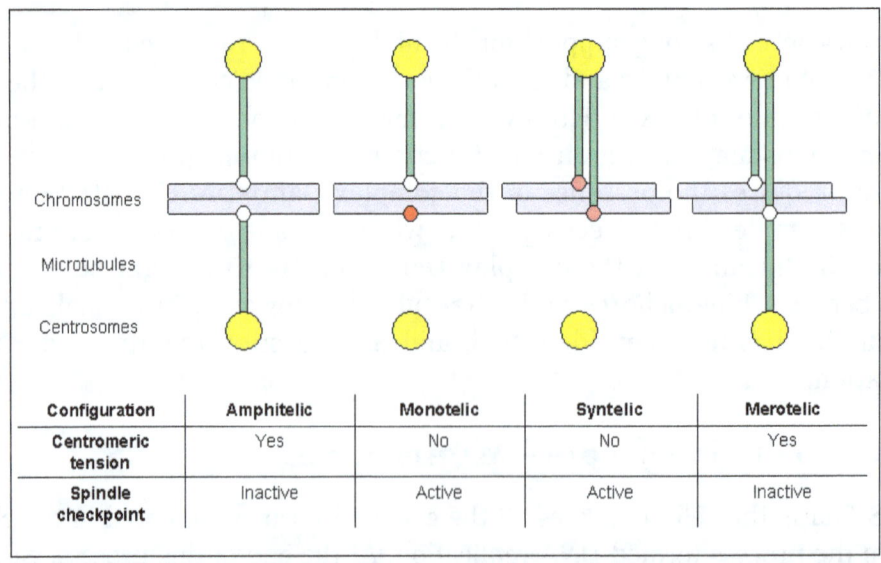

Configuration	Amphitelic	Monotelic	Syntelic	Merotelic
Centromeric tension	Yes	No	No	Yes
Spindle checkpoint	Inactive	Active	Active	Inactive

Different anchoring configurations between chromosomes and the mitotic spindle.

Both the monotelic and the syntelic configurations fail to generate centromeric tension and are detected by the spindle checkpoint. In contrast, the merotelic configuration is not detected by this control mechanism. However, most of these errors are detected and corrected before the cell enters in anaphase. A key factor in the correction of these anchoring errors is the chromosomal passenger complex, which includes the kinase protein Aurora B, its target and activating subunit INCENP and two other subunits, Survivin and Borealin/Dasra B. Cells in which the function of this complex has been abolished by dominant negative mutants, RNAi, antibody microinjection or using selective drugs, accumulate errors in chromosome anchoring. Many studies have shown that Aurora B is required to destabilize incorrect anchoring kinetochore-MT, favoring the generation of amphitelic connections. Aurora B homolog in yeast (Ipl1p) phosphorilates some kinetochore proteins, such as the constitutive protein Ndc10p and members of the Ndc80 and Dam1-DASH-DDD complexes. Phosphorylation of Ndc80 complex components produces destabilization of kMTs anchoring. It has been proposed that Aurora B localization is important for its function: as it is located in the inner region of the kinetochore (in the centromeric heterochromatin), when the centromeric tension is established sister kinetochores separate, and Aurora B cannot reach its substrates, so that kMTs are stabilized. Aurora B is frequently overexpressed in several cancer types, and it is currently a target for the development of anticancer drugs.

Spindle Checkpoint Activation

The spindle checkpoint, or SAC (for spindle assembly checkpoint), also known as the mitotic checkpoint, is a cellular mechanism responsible for detection of:

- Correct assembly of the mitotic spindle.

- Attachment of all chromosomes to the mitotic spindle in a bipolar manner.

- Congression of all chromosomes at the metaphase plate.

When just one chromosome (for any reason) remains lagging during congression, the spindle checkpoint machinery generates a delay in cell cycle progression: the cell is arrested, allowing time for repair mechanisms to solve the detected problem. After some time, if the problem has not been solved, the cell will be targeted for apoptosis (programmed cell death), a safety mechanism to avoid the generation of aneuploidy, a situation which generally has dramatic consequences for the organism.

Whereas structural centromeric proteins (such as CENP-B), remain stably localized throughout mitosis (including during telophase), the spindle checkpoint components are assembled on the kinetochore in high concentrations in the absence of microtubules, and their concentrations decrease as the number of microtubules attached to the kinetochore increases.

At metaphase, CENP-E, Bub3 and Bub1 levels decreases 3 to 4 fold as compared to the levels at unattached kinetochores, whereas the levels of dynein/dynactin, Mad1, Mad2 and BubR1 decrease >10-100 fold. Thus at metaphase, when all chromosomes are aligned at the metaphase plate, all checkpoint proteins are released from the kinetochore. The disappearance of the checkpoint proteins out of the kinetochores indicates the moment when the chromosome has reached the metaphase plate and is under bipolar tension. At this moment, the checkpoint proteins that bind to and inhibit Cdc20 (Mad1-Mad2 and BubR1), release Cdc20, which binds and activates APC/C^{Cdc20}, and this complex triggers sister chromatids separation and consequently anaphase entry.

Several studies indicate that the Ndc80 complex participates in the regulation of the stable association of Mad1-Mad2 and dynein with kinetochores. Yet the kinetochore associated proteins CENP-A, CENP-C, CENP-E, CENP-H and BubR1 are independent of Ndc80/Hec1. The prolonged arrest in prometaphase observed in cells with low levels of Ndc80/Hec1 depends on Mad2, although these cells show low levels of Mad1, Mad2 and dynein on kinetochores (<10-15% in relation to unattached kinetochores). However, if both Ndc80/Hec1 and Nuf2 levels are reduced, Mad1 and Mad2 completely disappear from the kinetochores and the spindle checkpoint is inactivated.

Shugoshin (Sgo1, MEI-S332 in Drosophila melanogaster) are centromeric proteins which are essential to maintain cohesin bound to centromeres until anaphase. The human homolog, hsSgo1, associates with centromeres during prophase and disappears when anaphase starts. When Shugoshin levels are reduced by RNAi in HeLa cells, cohesin cannot remain on the centromeres during mitosis, and consequently sister chromatids separate synchronically before anaphase initiates, which triggers a long mitotic arrest.

On the other hand, Dasso and collaborators have found that proteins involved in the Ran cycle can be detected on kinetochores during mitosis: RanGAP1 (a GTPase activating protein which stimulates the conversion of Ran-GTP in Ran-GDP) and the Ran binding protein called RanBP2/Nup358. During interphase, these proteins are located at the nuclear pores and participate in the nucleo-cytoplasmic transport. Kinetochore localization of these proteins seems to be functionally significant, because some treatments that increase the levels of Ran-GTP inhibit kinetochore release of Bub1, Bub3, Mad2 and CENP-E.

Orc2 (a protein that belongs to the origin recognition complex -ORC- implicated in DNA replication initiation during S phase) is also localized at kinetochores during mitosis in human cells; in agreement with this localization, some studies indicate that Orc2 in yeast is implicated in sister chromatids cohesion, and when it is eliminated from the cell, spindle checkpoint activation ensues. Some other ORC components (such orc5 in S. pombe) have been also found to participate in cohesion. However, ORC proteins seem to participate in a molecular pathway which is additive to cohesin pathway, and it is mostly unknown.

Force Generation to Propel Chromosome Movement

Most chromosome movements in relation to spindle poles are associated to lengthening and shortening of kMTs. One of the most interesting features of kinetochores is their capacity to modify the state of their associated kMTs (around 20) from a depolymerization state at their (+) end to polymerization state. This allows the kinetochores from cells at prometaphase to show "directional instability", changing between persistent phases of movement towards the pole (poleward) or inversed (anti-poleward), which are coupled with alternating states of kMTs depolymerization and polymerization, respectively. This kinetochore bi-stability seem to be part of a mechanism to align the chromosomes at the equator of the spindle without losing the mechanic connection between kinetochores and spindle poles. It is thought that kinetochore bi-stability is based upon the dynamic instability of the kMTs (+) end, and it is partially controlled by the tension present at the kinetochore. In mammalian cultured cells, a low tension at kinetochores promotes change towards kMTs depolymerization, and high tension promotes change towards kMTs polymerization.

Kinetochore proteins and proteins binding to MTs (+) end (collectively called +TIPs) regulate kinetochore movement through the kMTs (+) end dynamics regulation. However, the kinetochore-microtubule interface is highly dynamic, and some of these proteins seem to be bona fide components of both structures. Two groups of proteins seem to be particularly important: kinesins which work like depolymerases, such as KinI kinesins; and proteins bound to MT (+) ends, +TIPs, promoting polymerization, perhaps antagonizing the depolymerases effect:

- KinI kinesins are named "I" because they present an internal motor domain, which uses ATP to promote depolymerization of tubulin polymer, the microtubule. In vertebrates, the most important KinI kinesin controlling the dynamics of the (+) end assembly is MCAK. However, it seems that there are other kinesins implicated.

- There are two groups of +TIPs with kinetochore functions:

  ○ The first one includes the protein adenomatous polyposis coli (APC) and the associated protein EB1, which need MTs to localize on the kinetochores. Both proteins are required for correct chromosome segregation. EB1 binds only to MTs in polymerizing state, suggesting that it promotes kMTs stabilization during this phase.

  ○ The second group of +TIPs includes proteins which can localize on kinetochores even in absence of MTs. In this group there are two proteins that have been widely studied: CLIP-170 and their associated proteins CLASPs (CLIP-associated proteins). CLIP-170 role at kinetochores is unknown, but the expression of a dominant negative mutant produces a prometaphase delay, suggesting that it has an active role in chromosome alignment. CLASPs proteins are required for chromosome alignment and maintenance of a bipolar spindle in Drosophila, humans and yeast.

CELL GROWTH

The term cell growth is used in the contexts of biological cell development and cell division (reproduction). When used in the context of cell development, the term refers to increase in cytoplasmic and organelle volume (G_1 phase), as well as increase in genetic material (G_2 phase) following the replication during S phase. This is not to be confused with growth in the context of cell division, referred to as proliferation, where a cell, known as the "mother cell", grows and divides to produce two "daughter cells" (M phase).

Cell Populations

Cell populations go through a particular type of exponential growth called dowaiting. Thus, each generation of cells should be twice as numerous as the previous generation. However, the number of generations only gives a maximum figure as not all cells survive in each generation. Cells can reproduce in the stage of Mitosis, where they double and split into two genetically equal cells.

Measurement Methods

The cell growth can be detected by a variety of methods. The cell size growth can be visualized by microscopy, using suitable stains. But the increase of cells number is usually more significant. It can be measured by manual counting of cells under microscopy observation, using the dye exclusion method (i.e. trypan blue) to count only viable cells. Less fastidious, scalable, methods include the use of cytometers, while flow cytometry allows combining cell counts ('events') with other specific parameters: fluorescent probes for membranes, cytoplasm or nuclei allow distinguishing dead/viable cells, cell types, cell differentiation, expression of a biomarker such as Ki67.

Beside the increasing number of cells, one can be assessed regarding the metabolic activity growth, that is, the CFDA and calcein-AM measure (fluorimetrically) not only the membrane functionality (dye retention), but also the functionality of cytoplasmic enzymes (esterases). The MTT assays (colorimetric) and the resazurin assay (fluorimetric) dose the mitochondrial redox potential.

CELL DIVISION

Cell division is the process by which a parent cell divides into two or more daughter cells. Cell division usually occurs as part of a larger cell cycle. In eukaryotes, there are two distinct types of cell division: a vegetative division, whereby each daughter cell is genetically identical to the parent cell (mitosis), and a reproductive cell division, whereby the number of chromosomes in the daughter cells is reduced by half to produce haploid gametes (meiosis). Meiosis results in four haploid daughter cells by undergoing one round of DNA replication followed by two divisions. Homologous chromosomes are separated in the first division, and sister chromatids are separated in the second division. Both of these cell division cycles are used in the process of sexual reproduction at some point in their life cycle. Both are believed to be present in the last eukaryotic common ancestor.

Prokaryotes (bacteria) undergo a vegetative cell division known as binary fission, where their genetic material is segregated equally into two daughter cells. While binary fission may be the means of division by most prokaryotes, there are alternative manners of division, such as budding, that have been observed. All cell divisions, regardless of organism, are preceded by a single round of DNA replication.

For simple unicellular microorganisms such as the amoeba, one cell division is equivalent to reproduction – an entire new organism is created. On a larger scale, mitotic cell division can create progeny from multicellular organisms, such as plants that grow from cuttings. Mitotic cell division enables sexually reproducing organisms to develop from the one-celled zygote, which itself was produced by meiotic cell division from gametes. After growth, cell division by mitosis allows for continual construction and repair of the organism. The human body experiences about 10 quadrillion cell divisions in a lifetime.

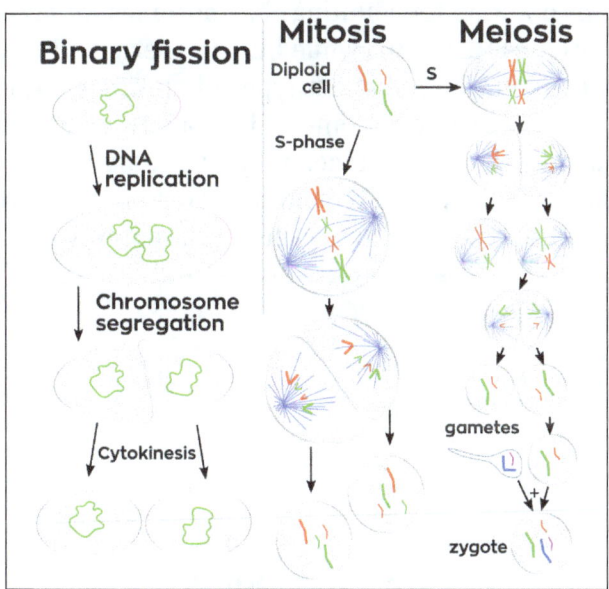

Three types of cell division.

The primary concern of cell division is the maintenance of the original cell's genome. Before division can occur, the genomic information that is stored in chromosomes must be replicated, and the duplicated genome must be separated cleanly between cells. A great deal of cellular infrastructure is involved in keeping genomic information consistent between generations.

Comparison of the Three Types of Cell Division

The DNA content of a cell is duplicated at the start of the cell reproduction process. Prior to DNA replication, the DNA content of a cell can be represented as the amount Z (the cell has Z chromosomes). After the DNA replication process, the amount of DNA in the cell is 2Z (multiplication: 2 x Z = 2Z). During Binary fission and mitosis the duplicated DNA content of the reproducing parental cell is separated into two equal halves that are destined to end up in the two daughter cells. The final part of the cell reproduction process is cell division, when daughter cells physically split apart from a parental cell. During meiosis, there are two cell division steps that together produce the four daughter cells.

After the completion of binary fission or cell reproduction involving mitosis, each daughter cell has the same amount of DNA (Z) as what the parental cell had before it replicated its DNA. These two types of cell reproduction produced two daughter cells that have the same number of chromosomes as the parental cell. Chromosomes duplicate prior to cell division when forming new skin cells for reproduction. After meiotic cell reproduction the four daughter cells have half the number of chromosomes that the parental cell originally had. This is the haploid amount of DNA, often symbolized as N. Meiosis is used by diploid organisms to produce haploid gametes. In a diploid organism such as the human organism, most cells of the body have the diploid amount of DNA, 2N. Using this notation for counting chromosomes we say that human somatic cells have 46 chromosomes (2N = 46) while human sperm and eggs have 23 chromosomes (N = 23). Humans have 23 distinct types of chromosomes, the 22 autosomes and the special category of sex chromosomes. There are two distinct sex chromosomes, the X chromosome and the Y chromosome. A diploid human cell has 23 chromosomes from that person's father and 23 from the mother. That is, your body has two copies of human chromosome number 2, one from each of your parents.

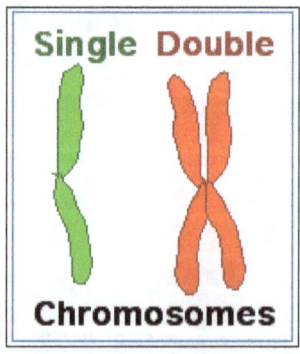

Chromosomes.

Immediately after DNA replication a human cell will have 46 "double chromosomes". In each double chromosome there are two copies of that chromosome's DNA molecule. During mitosis the double chromosomes are split to produce 92 "single chromosomes", half of which go into each daughter cell. During meiosis, there are two chromosome separation steps which assure that each of the four daughter cells gets one copy of each of the 23 types of chromosome.

Sexual Reproduction

Though cell reproduction that uses mitosis can reproduce eukaryotic cells, eukaryotes bother with the more complicated process of meiosis because sexual reproduction such as meiosis confers a selective advantage. Notice that when meiosis starts, the two copies of sister chromatids number 2 are adjacent to each other. During this time, there can be genetic recombination events. Information from the chromosome 2 DNA gained from one parent (red) will transfer over to the chromosome 2 DNA molecule that was received from the other parent (green). Notice that in mitosis the two copies of chromosome number 2 do not interact. Recombination of genetic information between homologous chromosomes during meiosis is a process for repairing DNA damages. This process can also produce new combinations of genes, some of which may be adaptively beneficial and influence the course of evolution. However, in organisms with more than one set of chromosomes at the main life cycle stage, sex may also provide an advantage because, under random mating, it produces homozygotes and heterozygotes according to the Hardy-Weinberg ratio.

Types of Cell Division

There are two distinct types of cell division out of which the first one is vegetative division, wherein each daughter cell duplicates the parent cell called mitosis. The second one is meiosis, which divides into four haploid daughter cells.

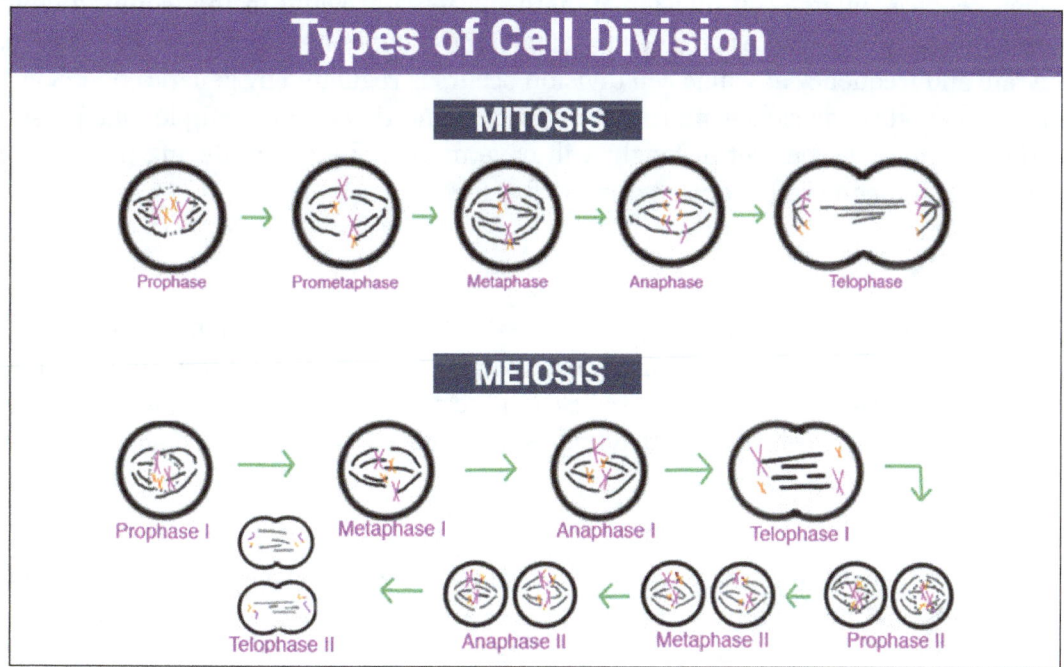

- Mitosis: The process cells use to make exact replicas of themselves. Mitosis is observed in almost all the body's cells, including eyes, skin, hair, and muscle cells.

- Meiosis: In this type of cell division, sperm or egg cells are produced instead of identical daughter cells as in mitosis.

- Binary Fission: Single-celled organisms like bacteria replicate themselves for reproduction.

Factors affecting Cell Division

Several factors cause and affect cell division. Some factors improve health and development while others cause cancer, birth defects, a variety of disorders and even death.

Nutrients

The nutrients present in the cell affect cell division. Certain nutrients such as vitamins, minerals and antioxidants are able to neutralize some chemicals in the body that cause cells to mutate and divide. Healthy nutrients obtained from consuming fruits and vegetables help to ensure that cells remain healthy and therefore cell division produces health cells. In the case of microorganisms, nutrients are absorbed from their surroundings.

Genetics

Genetic code regulates cell division. Whether a fetus growing in the womb, a child whose bones are growing or an elderly woman whose bones have begun to break down, the rate and frequency at which cell division occurs is regulated by genetic code. Some people's genetic code causes more cell division than others. For example, one person who grows to be seven feet in height will have more cell division during the growth phase than someone who stops growing at five feet.

Chemicals

Exposure to toxic chemicals such as pesticides and some cleaning chemicals can cause cell mutation. When cells mutate and then divide the results are multiple mutated and damaged cells. Mutated cells are the cause of illness and disease. Fortunately there are treatments to kill off cells that were damaged or mutated during cell division.

Stress

Stress affects cell division. Research shows that extreme stress levels can actually damage cells in the human body. If these cells are damaged yet still undergo cell division, the new cells will also be damaged. This can cause cancer and other diseases.

CYTOKINESIS

Cytokinesis is the part of the cell division process during which the cytoplasm of a single eukaryotic cell divides into two daughter cells. Cytoplasmic division begins during or after the late stages of nuclear division in mitosis and meiosis. During cytokinesis the spindle apparatus partitions and transports duplicated chromatids into the cytoplasm of the separating daughter cells. It thereby ensures that chromosome number and complement are maintained from one generation to the next and that, except in special cases, the daughter cells will be functional copies of the parent cell. After the completion of the telophase and cytokinesis, each daughter cell enters the interphase of the cell cycle.

Particular functions demand various deviations from the process of symmetrical cytokinesis; for example in oogenesis in animals the ovum takes almost all the cytoplasm and organelles. This leaves very little for the resulting polar bodies, which in most species die without function, though they do take on various special functions in other species. Another form of mitosis occurs in tissues such as liver and skeletal muscle; it omits cytokinesis, thereby yielding multinucleate cells.

Plant cytokinesis differs from animal cytokinesis, partly because of the rigidity of plant cell walls. Instead of plant cells forming a cleavage furrow such as develops between animal daughter cells, a dividing structure known as the cell plate forms in the cytoplasm and grows into a new, doubled cell wall between plant daughter cells. It divides the cell into two daughter cells.

Cytokinesis largely resembles the prokaryotic process of binary fission, but because of differences between prokaryotic and eukaryotic cell structures and functions, the mechanisms differ. For instance, a bacterial cell has only a single chromosome in the form of a closed loop, in contrast to the linear, usually multiple, chromosomes of eukaryote. Accordingly, bacteria construct no mitotic spindle in cell division. Also, duplication of prokaryotic DNA takes place during the actual separation of chromosomes; in mitosis, duplication takes place during the interphase before mitosis begins, though the daughter chromatids do not separate completely before the anaphase.

Animal Cell

Animal cell cytokinesis begins shortly after the onset of sister chromatid separation in the anaphase of mitosis. The process can be divided to the following distinct steps: anaphase spindle reorganization, division plane specification, actin-myosin ring assembly and contraction, and abscission. Faithful partitioning of the genome to emerging daughter cells is ensured through the tight temporal coordination of the above individual events by molecular signaling pathways.

Anaphase Spindle Reorganization

Animal cell cytokinesis starts with the stabilization of microtubules and reorganization of the mitotic spindle to form the central spindle. The central spindle (or spindle midzone) forms when non-kinetochore microtubule fibers are bundled between the spindle poles. A number of different species including H. sapiens, D. melanogaster and C. elegans require the central spindle in order to efficiently undergo cytokinesis, although the specific phenotype associated with it's absence varies from one species to the next (for example, certain Drosophila cell types are incapable of forming a cleavage furrow without the central spindle, whereas in both C. elegans embryos and human tissue culture cells a cleavage furrow is observed to form and ingress, but then regress before cytokinesis is complete). The process of mitotic spindle reorganization and central spindle formation is caused by the decline of CDK1 activity during anaphase. The decline of CDK1 activity at the metaphase-anaphase transition leads to dephosphorylating of inhibitory sites on multiple central spindle components. First of all, the removal of a CDK1 phosphorylation from a subunit of the CPC (the chromosomal passenger complex) allows its translocalization to the central spindle from the centromeres, where it is located during metaphase. Besides being a structural component of the central spindle itself, CPC also plays a role in the phosphoregulation of other central spindle components, including PRC1 (microtubule-bundling protein required for cytokinesis 1) and MKLP1 (a kinesin motor protein). Originally inhibited by CDK1-mediated phosphorylation, PRC1 is now able to form a homodimer that selectively binds to the interface between antiparallel microtubules, facilitating spatial organization of the microtubules of the central spindle. MKLP1, together with the Rho-family GTPase activating protein CYK-4 (also termed MgcRacGAP), forms the centralspindlin complex. Centralspindlin binds to the central spindle as higher-order clusters. The centralspindlin cluster formation is promoted by phosphorylation of MLKP1 by Aurora B, a component of CPC. In short, the self-assembly of central spindle is initiated through the phosphoregulation of multiple central spindle components by the decline of CDK1 activity, either directly or indirectly, at the metaphase-anaphase transition. The central spindle may have multiple functions in cytokinesis including the control of cleavage furrow positioning, the delivery of membrane vesicles to the cleavage furrow, and the formation of the midbody structure that is required for the final steps of division.

Division Plane Specification

The second step of animal cell cytokinesis involves division plane specification and cytokinetic furrow formation. Precise positioning of the division plane between the two masses of segregated chromosomes is essential to prevent chromosome loss. Meanwhile, the mechanism by which the spindle determines the division plane in animal cells is perhaps the most enduring mystery in cytokinesis and a matter of intense debate. There exist three hypotheses of furrow induction. The first is the astral stimulation hypothesis, which postulates that astral microtubules from the spindle poles carry a furrow-inducing signal to the cell cortex, where signals from two poles are somehow focused into a ring at the spindle. A second possibility, called the central spindle hypothesis, is that the cleavage furrow is induced by a positive stimulus that originates in the central spindle equator. The central spindle may contribute to the specification of the division plane by promoting concentration and activation of the small GTPase RhoA at the equatorial cortex. A third hypothesis is the astral relaxation hypothesis. It postulates that active actin-myosin bundles are distributed throughout the cell cortex, and inhibition of their contraction near the spindle poles results in a gradient of contractile activity that is highest at the midpoint between poles. In other words, astral microtubules generate a negative signal that increases cortical relaxation close to the poles. Genetic and laser-micromanipulation studies in C. elegans embryos have shown that the spindle sends two redundant signals to the cell cortex, one originating from the central spindle, and a second signal deriving from the spindle aster, suggesting the involvement of multiple mechanisms combined in the positioning of the cleavage furrow. The predominance of one particular signal varies between cell types and organisms. And the multitude and partial redundancy of signals may be required to make the system robust and to increase spatial precision.

Actin-myosin Ring Assembly and Contraction

At the cytokinesis furrow, it is the actin-myosin contractile ring that drives the cleavage process, during which cell membrane and wall grow inward, which eventually pinches the mother cell in two. The key components of this ring are the filamentous protein actin and the motor protein myosin II. The contractile ring assembles equatorially (in the middle of the cell) at the cell cortex (adjacent to the cell membrane). Rho protein family (RhoA protein in mammalian cells) is a key regulator of contractile ring formation and contraction in animal cells. The RhoA pathway promotes assembly of the actin-myosin ring by two main effectors. First, RhoA stimulates nucleation of unbranched actin filaments by activation of Diaphanous-related formins. This local generation of new actin filaments is important for the contractile ring formation. This actin filament formation process also requires a protein called profilin, which binds to actin monomers and helps load them onto the filament end. Second, RhoA promotes myosin II activation by the kinase ROCK, which activates myosin II directly by phosphorylation of the myosin light chain and also inhibits myosin phosphatase by phosphorylation of the phosphatase-targeting subunit MYPT. Besides actin and myosin II, the contractile ring contains the scaffolding protein anillin. Anillin binds to actin, myosin, RhoA, and CYK-4, and thereby links the equatorial cortex with the signals from the central spindle. It also contributes to the linkage of the actin-myosin ring to the plasma membrane. Another protein, septin, has also been speculated to serve as a structural scaffold on which the cytokinesis apparatus is organized. Following its assembly, contraction of the actin-myosin ring leads to ingression of the attached plasma membrane, which partitions the cytoplasm into two domains of emerging sister cells. The force for the contractile processes is generated by movements along actin by the motor protein myosin II. Myosin II uses the free energy released when ATP is hydrolyzed to move along these actin filaments, constricting the cell membrane to form a cleavage furrow. Continued hydrolysis causes this cleavage furrow to ingress (move inwards), a striking process that is clearly visible through a light microscope.

Abscission

The cytokinetic furrow ingresses until a midbody structure (composed of electron-dense, proteinaceous material) is formed, where the actin-myosin ring has reached a diameter of about 1–2 μm. Most animal cell types remain connected by an intercellular cytokinetic bridge for up to several hours until they are split by an actin-independent process termed abscission, the last step of cytokinesis. The process of abscission physically cleaves the midbody into two. Abscission proceeds by removal of cytoskeletal structures from the cytokinetic bridge, constriction of the cell cortex, and plasma membrane fission. The intercellular bridge is filled with dense bundles of antiparallel microtubules that derive from the central spindle. These microtubules overlap at the midbody, which is generally thought to be a targeting platform for the abscission machinery. Microtubule severing protein spastin is largely responsible for the disassembly of microtubule bundles inside the intercellular bridge. Complete cortical constriction also requires removal of the underlying cytoskeletal structures. Actin filament disassembly during late cytokinesis depends on the PKCε–14-3-3 complex, which inactivates RhoA after furrow ingression. Actin disassembly is further controlled by the GTPase Rab35 and its effector, the phosphatidylinositol-4,5-bisphosphate 5-phosphatase OCRL. Understanding the mechanism by which the plasma membrane ultimately splits requires further investigation.

Timing Cytokinesis

Cytokinesis must be temporally controlled to ensure that it occurs only after sister chromatids separate during the anaphase portion of normal proliferative cell divisions. To achieve this, many components of the cytokinesis machinery are highly regulated to ensure that they are able to perform a particular function at only a particular stage of the cell cycle. Cytokinesis happens only after APC binds with CDC20. This allows for the separation of chromosomes and myosin to work simultaneously. After cytokinesis, non-kinetochore microtubules reorganize and disappear into a new cytoskeleton as the cell cycle returns to interphase.

Cell Plate

Phragmoplast and cell plate formation in a plant cell during cytokinesis. Left side: Phragmoplast forms and cell plate starts to assemble in the center of the cell. Towards the right: Phragmoplast enlarges in a donut-shape towards the outside of the cell, leaving behind mature cell plate in the center. The cell plate will transform into the new cell wall once cytokinesis is complete.

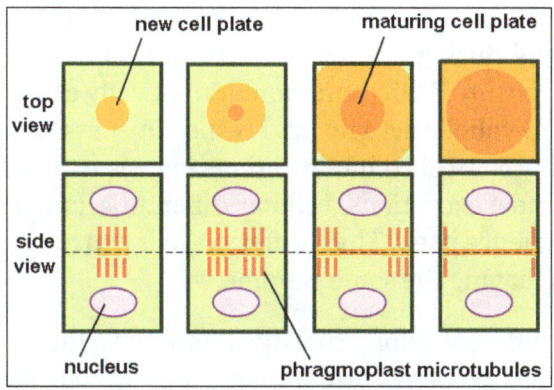

Cytokinesis in terrestrial plants occurs by cell plate formation. This process entails the delivery of Golgi-derived and endosomal vesicles carrying cell wall and cell membrane components to the plane of cell division and the subsequent fusion of these vesicles within this plate.

After formation of an early tubulo-vesicular network at the center of the cell, the initially labile cell plate consolidates into a tubular network and eventually a fenestrated sheet. The cell plate grows outward from the center of the cell to the parental plasma membrane with which it will fuse, thus completing cell division. Formation and growth of the cell plate is dependent upon the phragmoplast, which is required for proper targeting of Golgi-derived vesicles to the cell plate.

As the cell plate matures in the central part of the cell, the phragmoplast disassembles in this region and new elements are added on its outside. This process leads to a steady expansion of the phragmoplast and, concomitantly, to a continuous retargeting of Golgi-derived vesicles to the growing edge of the cell plate. Once the cell plate reaches and fuses with the plasma membrane the phragmoplast disappears. This event not only marks the separation of the two daughter cells, but also initiates a range of biochemical modifications that transform the callose-rich, flexible cell plate into a cellulose-rich, stiff primary cell wall.

The heavy dependence of cell plate formation on active Golgi stacks explains why plant cells, unlike animal cells, do not disassemble their secretion machinery during cell division.

CELL DIFFERENTIATION

Cellular differentiation is the process where a cell changes from one cell type to another. Usually, the cell changes to a more specialized type. Differentiation occurs numerous times during the development of a multicellular organism as it changes from a simple zygote to a complex system of tissues and cell types. Differentiation continues in adulthood as adult stem cells divide and create fully differentiated daughter cells during tissue repair and during normal cell turnover. Some differentiation occurs in response to antigen exposure. Differentiation dramatically changes a cell's size, shape, membrane potential, metabolic activity, and responsiveness to signals. These changes are largely due to highly controlled modifications in gene expression and are the study of epigenetics. With a few exceptions, cellular differentiation almost never involves a change in the DNA sequence itself. Thus, different cells can have very different physical characteristics despite having the same genome.

A specialized type of differentiation, known as 'terminal differentiation', is of importance in some tissues, for example vertebrate nervous system, striated muscle, epidermis and gut. During terminal differentiation, a precursor cell formerly capable of cell division, permanently leaves the cell cycle, dismantles the cell cycle machinery and often expresses a range of genes characteristic of the cell's final function (e.g. myosin and actin for a muscle cell). Differentiation may continue to occur after terminal differentiation if the capacity and functions of the cell undergo further changes.

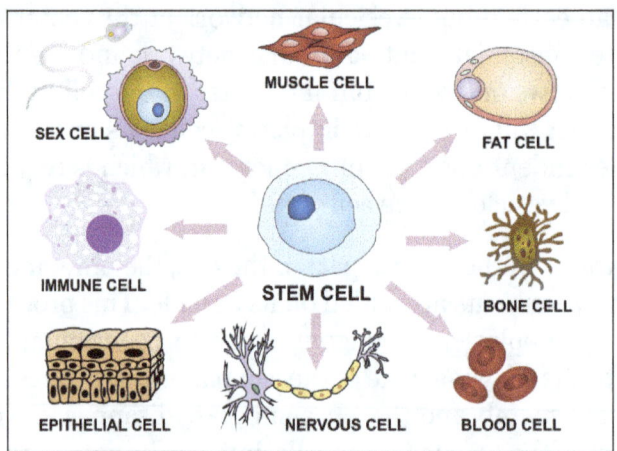

Stem cell differentiation into various tissue types.

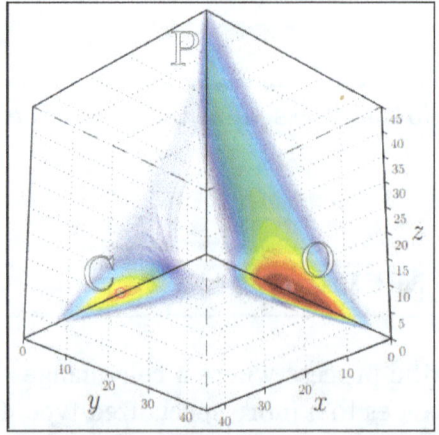

Cell-count distribution featuring cellular differentiation for three types of cells (progenitor x, osteoblast y and chondrocyte z) exposed to pro-osteoblast stimulus.

Among dividing cells, there are multiple levels of cell potency, the cell's ability to differentiate into other cell types. A greater potency indicates a larger number of cell types that can be derived. A cell that can differentiate into all cell types, including the placental tissue, is known as totipotent. In mammals, only the zygote and subsequent blastomeres are totipotent, while in plants, many differentiated cells can become totipotent with simple laboratory techniques. A cell that can differentiate into all cell types of the adult organism is known as pluripotent. Such cells are called meristematic cells in higher plants and embryonic stem cells in animals, though some groups report the presence of adult pluripotent cells. Virally induced expression of four transcription factors Oct4, Sox2, c-Myc, and Klf4 (Yamanaka factors) is sufficient to create pluripotent (iPS) cells from adult fibroblasts. A multipotent cell is one that can differentiate into multiple different, but closely related cell types. Oligopotent cells are more restricted than multipotent, but can still differentiate into a few closely related cell types. Finally, unipotent cells can differentiate into only one cell type, but are capable of self-renewal. In cytopathology, the level of cellular differentiation is used as a measure of cancer progression. "Grade" is a marker of how differentiated a cell in a tumor is.

Mammalian Cell Types

Three basic categories of cells make up the mammalian body: germ cells, somatic cells, and stem cells. Each of the approximately 37.2 trillion (3.72×10^{13}) cells in an adult human has its own copy or copies of the genome except certain cell types, such as red blood cells, that lack nuclei in their fully differentiated state. Most cells are diploid; they have two copies of each chromosome. Such cells, called somatic cells, make up most of the human body, such as skin and muscle cells. Cells differentiate to specialize for different functions.

Germ line cells are any line of cells that give rise to gametes—eggs and sperm—and thus are continuous through the generations. Stem cells, on the other hand, have the ability to divide for indefinite periods and to give rise to specialized cells. They are best described in the context of normal human development.

Development begins when a sperm fertilizes an egg and creates a single cell that has the potential to form an entire organism. In the first hours after fertilization, this cell divides into identical cells. In humans, approximately four days after fertilization and after several cycles of cell division, these cells begin to specialize, forming a hollow sphere of cells, called a blastocyst. The blastocyst has an outer layer of cells, and inside this hollow sphere, there is a cluster of cells called the inner cell mass. The cells of the inner cell mass go on to form virtually all of the tissues of the human body. Although the cells of the inner cell mass can form virtually every type of cell found in the human body, they cannot form an organism. These cells are referred to as pluripotent.

Pluripotent stem cells undergo further specialization into multipotent progenitor cells that then give rise to functional cells. Examples of stem and progenitor cells include:

- Radial glial cells (embryonic neural stem cells) that give rise to excitatory neurons in the fetal brain through the process of neurogenesis.

- Hematopoietic stem cells (adult stem cells) from the bone marrow that give rise to red blood cells, white blood cells, and platelets.

- Mesenchymal stem cells (adult stem cells) from the bone marrow that give rise to stromal cells, fat cells, and types of bone cells.

- Epithelial stem cells (progenitor cells) that give rise to the various types of skin cells.

- Muscle satellite cells (progenitor cells) that contribute to differentiated muscle tissue.

A pathway that is guided by the cell adhesion molecules consisting of four amino acids, arginine, glycine, asparagine, and serine, is created as the cellular blastomere differentiates from the single-layered blastula to the three primary layers of germ cells in mammals, namely the ectoderm, mesoderm and endoderm (listed from most distal (exterior) to proximal (interior)). The ectoderm ends up forming the skin and the nervous system, the mesoderm forms the bones and muscular tissue, and the endoderm forms the internal organ tissues.

Dedifferentiation

Dedifferentiation, or integration is a cellular process often seen in more basal life forms such as worms and amphibians in which a partially or terminally differentiated cell reverts to an earlier developmental stage, usually as part of a regenerative process. Dedifferentiation also occurs in plants. Cells in cell culture can lose properties they originally had, such as protein expression, or change shape. This process is also termed dedifferentiation.

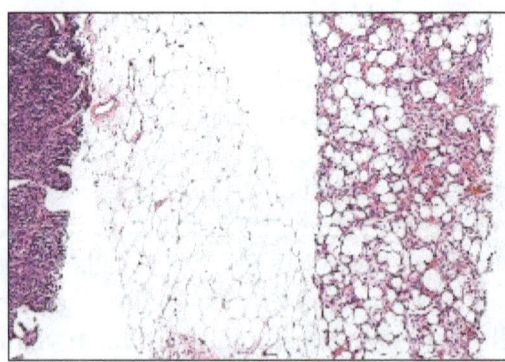

Micrograph of a liposarcoma with some dedifferentiation, that is not identifiable as a liposarcoma, (left edge of image) and a differentiated component (with lipoblasts and increased vascularity (right of image)). Fully differentiated (morphologically benign) adipose tissue (center of the image) has few blood vessels.

Some believe dedifferentiation is an aberration of the normal development cycle that results in cancer, whereas others believe it to be a natural part of the immune response lost by humans at some point as a result of evolution. A small molecule dubbed reversine, a purine analog, has been discovered that has prov-en to induce dedifferentiation in myotubes. These dedifferentiated cells could then re-differentiate into osteoblasts and adipocytes.

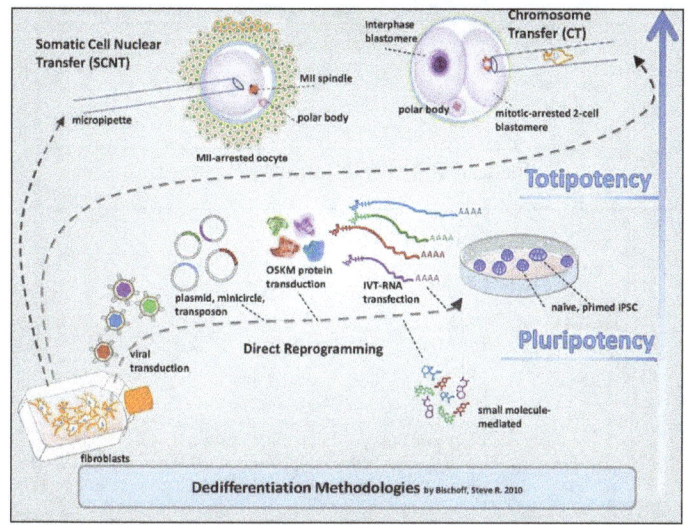

Several methods used to revert adult somatic cells to totipotency or pluripotency.

Mechanisms

Each specialized cell type in an organism expresses a subset of all the genes that con-stitute the genome of that species. Each cell type is defined by its particular pattern of regulated gene expression. Cell differentiation is thus a transition of a cell from one cell type to another and it involves a switch from one pattern of gene expression to another. Cellular differentiation during development can be understood as the result of a gene regulatory network. A regulatory gene and its cis-regulatory modules are nodes in a gene regulatory network; they receive input and create output elsewhere in the net-work. The systems biology approach to developmental biology emphasizes the impor-tance of investigating how developmental mechanisms interact to produce predictable patterns (morphogenesis). However, an alternative view has been proposed recently. Based on stochastic gene expression, cellular differentiation is the result of a Darwinian selective process occurring among cells. In this frame, protein and gene networks are the result of cellular processes and not their cause.

While evolutionarily conserved molecular processes are involved in the cellular mech-anisms underlying these switches, in animal species these are very different from the well-characterized gene regulatory mechanisms of bacteria, and even from those of the

animals' closest unicellular relatives. Specifically, cell differentiation in animals is highly dependent on biomolecular condensates of regulatory proteins and enhancer DNA
sequences.

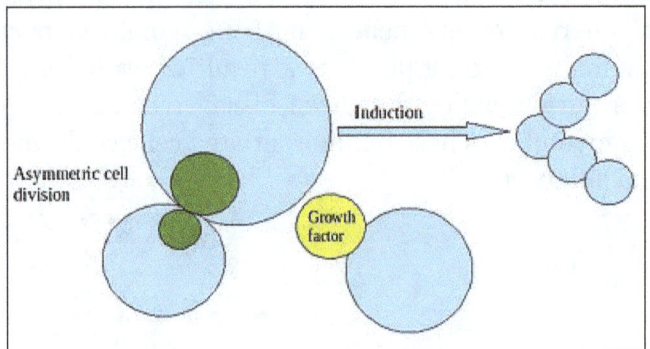

Mechanisms of cellular differentiation.

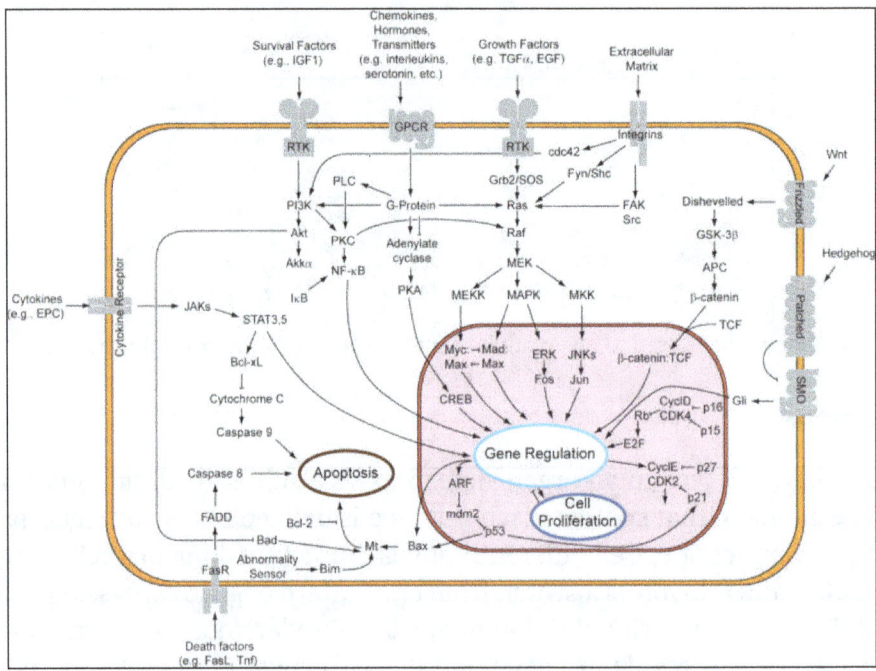

An overview of major signal transduction pathways.

Cellular differentiation is often controlled by cell signaling. Many of the signal molecules
that convey information from cell to cell during the control of cellular differentiation
are called growth factors. Although the details of specific signal transduction pathways
vary, these pathways often share the following general steps. A ligand produced by one
cell binds to a receptor in the extracellular region of another cell, inducing a conformational change in the receptor. The shape of the cytoplasmic domain of the receptor
changes, and the receptor acquires enzymatic activity. The receptor then catalyzes reactions that phosphorylate other proteins, activating them. A cascade of phosphorylation
reactions eventually activates a dormant transcription factor or cytoskeletal protein,

thus contributing to the differentiation process in the target cell. Cells and tissues can vary in competence, their ability to respond to external signals.

Signal induction refers to cascades of signaling events, during which a cell or tissue signals to another cell or tissue to influence its developmental fate. Yamamoto and Jeffery investigated the role of the lens in eye formation in cave- and surface-dwelling fish, a striking example of induction. Through reciprocal transplants, Yamamoto and Jeffery found that the lens vesicle of surface fish can induce other parts of the eye to develop in cave- and surface-dwelling fish, while the lens vesicle of the cave-dwelling fish cannot.

Other important mechanisms fall under the category of asymmetric cell divisions, divisions that give rise to daughter cells with distinct developmental fates. Asymmetric cell divisions can occur because of asymmetrically expressed maternal cytoplasmic determinants or because of signaling. In the former mechanism, distinct daughter cells are created during cytokinesis because of an uneven distribution of regulatory molecules in the parent cell; the distinct cytoplasm that each daughter cell inherits results in a distinct pattern of differentiation for each daughter cell. A well-studied example of pattern formation by asymmetric divisions is body axis patterning in Drosophila. RNA molecules are an important type of intracellular differentiation control signal. The molecular and genetic basis of asymmetric cell divisions has also been studied in green algae of the genus Volvox, a model system for studying how unicellular organisms can evolve into multicellular organisms. In Volvox carteri, the 16 cells in the anterior hemisphere of a 32-cell embryo divide asymmetrically, each producing one large and one small daughter cell. The size of the cell at the end of all cell divisions determines whether it becomes a specialized germ or somatic cell.

Epigenetic Control

Since each cell, regardless of cell type, possesses the same genome, determination of cell type must occur at the level of gene expression. While the regulation of gene expression can occur through cis- and trans-regulatory elements including a gene's promoter and enhancers, the problem arises as to how this expression pattern is maintained over numerous generations of cell division. As it turns out, epigenetic processes play a crucial role in regulating the decision to adopt a stem, progenitor, or mature cell fate.

In systems biology and mathematical modeling of gene regulatory networks, cell-fate determination is predicted to exhibit certain dynamics, such as attractor-convergence (the attractor can be an equilibrium point, limit cycle or strange attractor) or oscillatory.

Mechanisms of Epigenetic Regulation

Pioneer Factor Pioneering Factors (Oct$_4$, Sox$_2$, Nanog)

Three transcription factors, OCT4, SOX2, and NANOG – the first two of which are used in induced pluripotent stem cell (iPSC) reprogramming, along with Klf4 and c-Myc – are highly expressed in undifferentiated embryonic stem cells and are necessary for the maintenance of their pluripotency. It is thought that they achieve this through alterations in chromatin structure, such as histone modification and DNA methylation, to restrict or permit the transcription of target genes. While highly expressed, their levels require a precise balance to maintain pluripotency, perturbation of which will promote differentiation towards different lineages based on how the gene expression levels change. Differential regulation of Oct-4 and SOX2 levels have been shown to precede germ layer fate selection. Increased levels of Oct4 and decreased levels of Sox2 promote a mesendodermal fate, with Oct4 actively suppressing genes associated with a neural ectodermal fate. Similarly, Increased levels of Sox2 and decreased levels of Oct4 promote differentiation towards a neural ectodermal fate, with Sox2 inhibiting differentiation towards a mesendodermal fate. Regardless of the lineage cells differentiate down, suppression of NANOG has been identified as a necessary prerequisite for differentiation.

Polycomb Repressive Complex (PRC2)

In the realm of gene silencing, Polycomb repressive complex 2, one of two classes of the Polycomb group (PcG) family of proteins, catalyzes the di- and tri-methylation of histone H3 lysine 27 (H3K27me2/me3). By binding to the H3K27me2/3-tagged nucleosome, PRC1 (also a complex of PcG family proteins) catalyzes the mono-ubiquitinylation of histone H2A at lysine 119 (H2AK119Ub1), blocking RNA polymerase II activity and resulting in transcriptional suppression. PcG knockout ES cells do not differentiate efficiently into the three germ layers, and deletion of the PRC1 and PRC2 genes leads to increased expression of lineage-affiliated genes and unscheduled differentiation. Presumably, PcG complexes are responsible for transcriptionally repressing differentiation and development-promoting genes.

Trithorax Group Proteins (TrxG)

Alternately, upon receiving differentiation signals, PcG proteins are recruited to promoters of pluripotency transcription factors. PcG-deficient ES cells can begin differentiation but cannot maintain the differentiated phenotype. Simultaneously, differentiation and development-promoting genes are activated by Trithorax group (TrxG) chromatin regulators and lose their repression. TrxG proteins are recruited at regions of high transcriptional activity, where they catalyze the trimethylation of histone H3 lysine 4 (H3K4me3) and promote gene activation through histone acetylation. PcG and TrxG complexes engage in direct competition and are thought to be functionally antagonistic, creating at differentiation and development-promoting loci what is termed a "bivalent domain" and rendering these genes sensitive to rapid induction or repression.

DNA Methylation

Regulation of gene expression is further achieved through DNA methylation, in which the DNA methyltransferase-mediated methylation of cytosine residues in CpG dinucleotides maintains heritable repression by controlling DNA accessibility. The majority of CpG sites in embryonic stem cells are unmethylated and appear to be associated with H3K4me3-carrying nucleosomes. Upon differentiation, a small number of genes, including OCT4 and NANOG, are methylated and their promoters repressed to prevent their further expression. Consistently, DNA methylation-deficient embryonic stem cells rapidly enter apoptosis upon in vitro differentiation.

Nucleosome Positioning

While the DNA sequence of most cells of an organism is the same, the binding patterns of transcription factors and the corresponding gene expression patterns are different. To a large extent, differences in transcription factor binding are determined by the chromatin accessibility of their binding sites through histone modification and pioneer factors. In particular, it is important to know whether a nucleosome is covering a given genomic binding site or not. This can be determined using a chromatin immunoprecipitation (ChIP) assay.

Histone Acetylation and Methylation

DNA-nucleosome interactions are characterized by two states: either tightly bound by nucleosomes and transcriptionally inactive, called heterochromatin, or loosely bound and usually, but not always, transcriptionally active, called euchromatin. The epigenetic processes of histone methylation and acetylation, and their inverses demethylation and deacetylation primarily account for these changes. The effects of acetylation and deacetylation are more predictable. An acetyl group is either added to or removed from the positively charged Lysine residues in histones by enzymes called histone acetyltransferases or histone deacteylases, respectively. The acetyl group prevents Lysine's association with the negatively charged DNA backbone. Methylation is not as straightforward, as neither methylation nor demethylation consistently correlate with either gene activation or repression. However, certain methylations have been repeatedly shown to either activate or repress genes. The trimethylation of lysine 4 on histone 3 (H3K4Me3) is associated with gene activation, whereas trimethylation of lysine 27 on histone 3 represses genes.

In Stem Cells

During differentiation, stem cells change their gene expression profiles. Recent studies have implicated a role for nucleosome positioning and histone modifications during this process. There are two components of this process: turning off the expression of embryonic stem cell (ESC) genes, and the activation of cell fate genes. Lysine specific demethylase 1 (KDM1A) is thought to prevent the use of enhancer regions of pluripotency genes, thereby inhibiting their transcription. It interacts with Mi-2/NuRD complex (nucleosome remodelling and histone deacetylase) complex, giving an instance where methylation and acetylation are not discrete and mutually exclusive, but intertwined processes.

Role of Signaling in Epigenetic Control

The first major candidate is Wnt signaling pathway. The Wnt pathway is involved in all stages of differentiation, and the ligand Wnt3a can substitute for the overexpression of c-Myc in the generation of induced pluripotent stem cells. On the other hand, disruption of ß-catenin, a component of the Wnt signaling pathway, leads to decreased proliferation of neural progenitors.

Growth factors comprise the second major set of candidates of epigenetic regulators of cellular differentiation. These morphogens are crucial for development, and include bone morphogenetic proteins, transforming growth factors (TGFs), and fibroblast growth factors (FGFs). TGFs and FGFs have been shown to sustain expression of OCT4, SOX2, and NANOG by downstream signaling to Smad proteins. Depletion of growth factors promotes the differentiation of ESCs, while genes with bivalent chromatin can become either more restrictive or permissive in their transcription.

Several other signaling pathways are also considered to be primary candidates. Cytokine leukemia inhibitory factors are associated with the maintenance of mouse ESCs in an undifferentiated state. This is achieved through its activation of the Jak-STAT3 pathway, which has been shown to be necessary and sufficient towards maintaining mouse ESC pluripotency. Retinoic acid can induce differentiation of human and mouse ESCs, and Notch signaling is involved in the proliferation and self-renewal of stem cells. Finally, Sonic hedgehog, in addition to its role as a morphogen, promotes embryonic stem cell differentiation and the self-renewal of somatic stem cells.

The problem, of course, is that the candidacy of these signaling pathways was inferred primarily on the basis of their role in development and cellular differentiation. While epigenetic regulation is necessary for driving cellular differentiation, they are certainly not sufficient for this process. Direct modulation of gene expression through modification of transcription factors plays a key role that must be distinguished from heritable epigenetic changes that can persist even in the absence of the original environmental signals. Only a few examples of signaling pathways leading to epigenetic changes that alter cell fate currently exist, and we will focus on one of them.

Expression of Shh (Sonic hedgehog) upregulates the production of BMI1, a component of the PcG complex that recognizes H3K27me3. This occurs in a Gli-dependent manner, as Gli1 and Gli2 are downstream effectors of the Hedgehog signaling pathway. In culture, Bmi1 mediates the Hedgehog pathway's ability to promote human mammary stem cell self-renewal. In both humans and mice, researchers showed Bmi1 to be highly expressed in proliferating immature cerebellar granule cell precursors. When Bmi1 was knocked out in mice, impaired cerebellar development resulted, leading to significant reductions in postnatal brain mass along with abnormalities in motor control and behavior. A separate study showed a significant decrease in neural stem cell proliferation along with increased astrocyte proliferation in Bmi null mice.

An alternative model of cellular differentiation during embryogenesis is that positional information is based on mechanical signalling by the cytoskeleton using Embryonic differentiation waves. The mechanical signal is then epigenetically transduced via signal transduction systems (of which specific molecules such as Wnt are part) to result in differential gene expression.

In summary, the role of signaling in the epigenetic control of cell fate in mammals is largely unknown, but distinct examples exist that indicate the likely existence of further such mechanisms.

Effect of Matrix Elasticity

In order to fulfill the purpose of regenerating a variety of tissues, adult stems are known to migrate from their niches, adhere to new extracellular matrices (ECM) and differentiate. The ductility of these microenvironments are unique to different tissue types. The ECM surrounding brain, muscle and bone tissues range from soft to stiff. The transduction of the stem cells into these cells types is not directed solely by chemokine cues and cell to cell signaling. The elasticity of the microenvironment can also affect the differentiation of mesenchymal stem cells (MSCs which originate in bone marrow.) When MSCs are placed on substrates of the same stiffness as brain, muscle and bone ECM, the MSCs take on properties of those respective cell types. Matrix sensing requires the cell to pull against the matrix at focal adhesions, which triggers a cellular mechano-transducer to generate a signal to be informed what force is needed to deform the matrix. To determine the key players in matrix-elasticity-driven lineage specification in MSCs, different matrix microenvironments were mimicked. From these experiments, it was concluded that focal adhesions of the MSCs were the cellular mechano-transducer sensing the differences of the matrix elasticity. The non-muscle myosin IIa-c isoforms generates the forces in the cell that lead to signaling of early commitment markers. Nonmuscle myosin IIa generates the least force increasing to non-muscle myosin IIc. There are also factors in the cell that inhibit non-muscle myosin II, such as blebbistatin. This makes the cell effectively blind to the surrounding matrix. Researchers have obtained some success in inducing stem cell-like properties in HEK 239 cells by providing a soft matrix without the use of diffusing factors. The stem-cell properties appear to be linked to tension in the cells' actin network. One identified mechanism for matrix-induced differentiation is tension-induced proteins, which remodel chromatin in response to mechanical stretch. The RhoA pathway is also implicated in this process.

Factors Involved in Cell Differentiation

During cell differentiation in multicellular organisms, cells become specialized and take on roles such as those of nerve, muscle and blood cells. Factors involved in triggering cell differentiation include cell signaling, environmental influences and the level of development of the organism.

Basic cell differentiation occurs after a sperm cell fertilizes an egg and the resulting zygote reaches a certain size. At that point the zygote starts developing different cell types and needs differentiated cells to take on the specialized functions.

The mechanism that is at the root of cell differentiation is gene expression. All the cells of an organism have identical sets of genes because the genetic code was copied from the original egg cell fertilized by the sperm cell. To take on a specialized function, a cell will only express or use some of the genes in its genetic code and ignore the rest.

For example, a cell that differentiates to become a liver cell will express the liver cell genes, and all the other liver cells will use the same set of liver genes. They will differentiate together to form the liver.

Cell differentiation takes place in three situations:

- The growth of an immature organism into an adult.

- Normal turnover of cells such as blood cells in mature organisms.

- The repair of damaged tissues when specialized cells have to be replaced.

In each case, cell signaling informs cells what type of specialized cell is required. Undifferentiated cells express the corresponding genes to fulfill the needs of the organism.

Gene Expression Works by making Copies of the Gene

The genetic code of eukaryotic cells is located on the DNA in the nucleus. The DNA can't leave the nucleus so the cell has to copy the gene it wants to express.

Messenger RNA (mRNA) attaches to the DNA and copies the relevant gene. The mRNA can travel outside the nucleus and bring the genetic instructions to ribosomes that are floating in the cell cytoplasm or that are attached to the endoplasmic reticulum. The ribosomes produce the protein encoded by the expressed gene.

Depending on the signals received by the cell, the environmental influences and the developmental stage of the cell, the process of gene expression can be blocked at any stage. If the protein encoded by the gene is not needed by the organism, the mRNA will not copy the gene, and the gene expression process will not start.

Even after the mRNA copies the gene, the mRNA molecule may be blocked from exiting the nucleus or may not be able to reach a ribosome. Ribosomes may not produce the required protein even if mRNA delivers the copied genetic code. Different factors can influence gene expression all through this multi-step process.

Internal Factors that affect Cell Specialization

Organisms have several ways of ensuring that cells develop into the specialized and differentiated cells needed. The key factor driving cellular differentiation in the body is the manufacture of proteins. Cells can differentiate depending on which genes are expressed and which proteins are encoded in the expressed genes. The produced proteins help the differentiated cells perform their specialized function and let them tell other cells what they are doing through cell signaling.

A further mechanism that can influence cell differentiation is asymmetric segregation in cell division. Substances such as special proteins gather at one end of a cell. When the cell divides, one daughter cell has more of the special proteins than the other. The cells become different types of cells due to the different protein distribution.

As a cell differentiates, the type of specialization it can take on becomes more limited. Embryonic stem cells can initially become any type of cell, but once the cell is mature and has taken on a specialized role, it often can no longer change. Embryonic stem cells are called totipotent cells because they can still take on any role while mature, specialized cells that are fully differentiated can only carry out their specialized function.

Asymmetric Segregation Produces Different Cells

Gene expression is responsible for cell specialization, but the basic cells have to be able to take on the specialized functions. Before differentiation and cell specialization can take place, the right type of cell has to be available. Asymmetric segregation can produce such different types of cells. Totipotent embryonic cells become one of three types of pluripotent cells that eventually differentiate into the various body tissues.

The three types of pluripotent cells are:

- Endoderm cells become the lining of the respiratory and digestive tracts as well as forming the liver and many of the major glands such as the pancreas.

- Mesoderm cells differentiate to form muscles, bones, connective tissue and the heart.

- Ectoderm cells form the skin and nerves.

While cell signaling is responsible for the production of some different cell types and for cell specialization, asymmetric segregation acts at the beginning of cell development to produce pluripotent cells.

DNA transcription to mRNA takes place in such as way that the mRNA produces certain proteins at one end of the cell and different proteins at the other end. Cell division results in two different types of daughter cells that can go on to produce cells with different specializations.

Cell Signaling is at the Root of Cell Differentiation

Internal mechanisms that influence the cell differentiation of pluripotent cells are mainly based on cell signaling. Cells receive chemical signals that tell them what type of cell or what kind of protein is needed.

Cell signaling mechanisms include:

- Diffusion, in which cells release chemicals that spread throughout the tissues.

- Direct contact, in which cells have special chemicals on their cell membranes.

- Gap junctions, in which signaling chemicals can pass directly from one cell to another.

Cells continuously send out chemical messages regarding their activities and receive signals about what is going on in their immediate neighborhood, in the tissues where they are located and in the body at large. These signals are the principal factors that affect cell specialization, and cell signaling is the key factor driving cell differentiation in the body.

Cell Signaling by Diffusion Influences Tissue Development

Cells become sensitive to certain chemical signals because they have receptors on their cell membrane. The receptors depend on the type of cell, how it has developed and which genes are being expressed. As receptors are activated, the cell differentiates further.

When a cell sends a signal to many nearby cells, it emits a chemical that diffuses through the tissue in which the cell is embedded. The chemical signal is captured by receptors in the cell membranes of the surrounding cells and triggers a response inside each cell. These responses help cause the cells to differentiate in a way that builds tissue.

For example, cells that will become part of a liver emit chemicals that trigger the corresponding receptors in nearby cells, and all the cells in that location differentiate to become liver cells. As the liver tissue forms, further cell signaling triggers some cells to differentiate into duct cells or connecting tissue. Eventually the differentiated cells form a complete and functional liver.

Local Cell Signaling lets Cells Recognize their Neighbors

To develop into the specialized cells needed by the organism, cells have to know what other cells in their immediate surroundings are doing. Special receptors for cell-to-cell contact and gap junctions between cells facilitate the direct exchange of signals between neighboring cells. Cells can ensure that their surroundings correspond to their differentiated specialization.

In cell-to-cell signaling, specially formed receptor proteins on the surface of a cell match corresponding proteins on a neighboring cell's membrane. When the cells come into contact, the two proteins link, and a signal is triggered from one cell to the other.

The signal passes through the cell membrane and enters the cell where it causes a specific cell behavior.

For example, skin cells have to make sure they have other skin cells around them, but some skin cells will have the cells of the underlying tissue beneath them. Cell-to-cell signaling lets cells ensure that their surroundings match their differentiation.

Gap junctions are special links between neighboring cells that allow an easy and direct exchange of proteins acting as messages. Using gap junctions, cells can coordinate their activities and exchange signals quickly and easily.

For example, nerve cells use gap junctions to establish nerve pathways, and gap junctions let cells differentiate into the type of nerve cell that is appropriate to their location in the skin, in the spinal cord or in the brain.

Factors affecting Cell Signaling Influence Cell Differentiation

Cell signaling and the resulting cell differentiation are complex processes with many steps. Signals have to be produced, propagated received and acted upon. Triggers resulting from cell signals have to work as expected. Factors that disrupt any of the steps can influence cell differentiation and cause changes in the organism.

Factors that can influence and disrupt cell signaling and cell differentiation include a lack of nutrients; if a cell can't produce a protein because it lacks the building blocks, it can't differentiate. Mutations in the genetic code are another problem.

If the DNA is defective or the transcription is wrong, the signaling and differentiation process is disrupted. In addition to these, if the signaling chemicals are blocked or the cell receptors are filled with non-signaling chemical bonds, the signaling process will not work properly.

Environmental Factors can Influence Cell Differentiation

Influences from the environment of the organism that can affect cell signaling, gene expression and cell differentiation can change, stop or disrupt the process. Some environmental factors are used by the organism for adaptation, some can be used to fight disease and some harm or kill the organism.

For example, environmental temperature can influence the development of some organisms. Higher temperatures speed up the growth of cells and their differentiation while low temperatures slow down or stop development.

Drugs can disrupt harmful cell differentiation. For example, drugs can block one of the process steps for unlimited tumor growth and stop the expression of the corresponding genes.

Injuries can affect gene expression and influence what type of cell is needed to repair damage. Viruses and bacteria can influence cell differentiation. For example, if a mother is infected with a disease such as rubella, the developing fetus can have its cell differentiation influenced, and it can develop birth defects.

Finally toxic chemicals can affect cell differentiation. Substances that attack or block signaling chemicals or that block signal receptor positions on cell membranes can stop signaling activity and influence cell differentiation.

In the case of these environmental factors, the organism tries to respond by adapting or by changing internal processes. Adaptation is effective for some of the environmental influences, but for others, the organism may survive but exhibit defects, or the organism may die.

References

- Fukada S, Uezumi A, Ikemoto M, Masuda S, Segawa M, Tanimura N, Yamamoto H, Miyagoe-Suzuki Y, Takeda S (October 2007). "Molecular signature of quiescent satellite cells in adult skeletal muscle". Stem Cells. 25 (10): 2448–59. doi:10.1634/stemcells.2007-0019. PMID 17600112

- David M (2007). The cell cycle : principles of control. Oxford University Press. ISBN 978-0199206100. OCLC 813540567

- Factors-involved-cell-differentiation-6935462: sciencing.com, Retrieved 23 January, 2020

- Alberts B, Johnson A, Lewis J, Raff M, Roberts K, Walter P (2008). "Chapter 18 Apoptosis: Programmed Cell Death Eliminates Unwanted Cells". Molecular Biology of the Cell (textbook) (5th ed.). Garland Science. p. 1115. ISBN 978-0-8153-4105-5

- Shiloh Y, Kastan MB (2001). "ATM: genome stability, neuronal development, and cancer cross paths". Advances in Cancer Research. 83: 209–54. doi:10.1016/s0065-230x(01)83007-4. ISBN 9780120066834. PMID 11665719

- Blagosklonny MV, Pardee AB (2002). "The restriction point of the cell cycle". Cell Cycle. 1 (2): 103–10. doi:10.4161/cc.1.2.108. PMID 12429916

- Cell-cycle, cell-growth-death, basics: teachmephysiology.com, Retrieved 11 April, 2020

- Tsonis PA (April 2004). "Stem cells from differentiated cells". Mol. Interv. 4 (2): 81–3. doi:10.1124/mi.4.2.4. PMID 15087480. Archived from the original on 2016-05-23. Retrieved 2010-12-26

- What-is-a-cell-14023083, scitable: nature.com, Retrieved 20 March, 2020

- El-Naggar, A.K.; Chan, J.C.K.; Grandis, J.R.; Takata, T.; Slootweg, P.J. (2017-01-23). WHO Classification of Head and Neck Tumours. Lyon: International Agency for Research on Cancer. ISBN 978-92-832-2438-9

3

Phases in Mitosis

The process of mitosis mainly consists of five major phases to undergo genetically identical cell division. It includes prophase, prometaphase, metaphase, anaphase and telophase. This chapter closely examines these phases of mitosis to provide an extensive understanding of the subject.

Mitosis is a process of cell duplication, or reproduction, during which one cell gives rise to two genetically identical daughter cells. Strictly applied, the term mitosis is used to describe the duplication and distribution of chromosomes, the structures that carry the genetic information.

Prior to the onset of mitosis, the chromosomes have replicated and the proteins that will form the mitotic spindle have been synthesized. Mitosis begins at prophase with the thickening and coiling of the chromosomes. The nucleolus, a rounded structure, shrinks and disappears. The end of prophase is marked by the beginning of the organization of a group of fibres to form a spindle and the disintegration of the nuclear membrane.

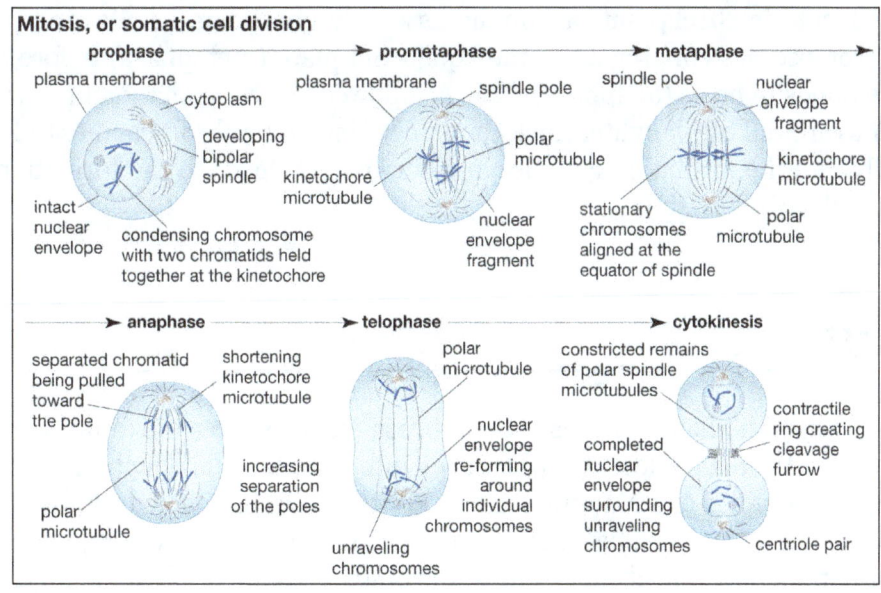

The process of cell division by mitosis.

The chromosomes, each of which is a double structure consisting of duplicate chromatids, line up along the midline of the cell at metaphase. In anaphase each chromatid pair separates into two identical chromosomes that are pulled to opposite ends of the cell by the spindle fibres. During telophase, the chromosomes begin to decondense, the spindle breaks down, and the nuclear membranes and nucleoli re-form. The cytoplasm of the mother cell divides to form two daughter cells, each containing the same number and kind of chromosomes as the mother cell. The stage, or phase, after the completion of mitosis is called interphase.

Mitosis is absolutely essential to life because it provides new cells for growth and for replacement of worn-out cells. Mitosis may take minutes or hours, depending upon the kind of cells and species of organisms. It is influenced by time of day, temperature, and chemicals.

M Phase

The M phase consists of mitosis and cytokinesis. Mitosis is the process in which DNA condenses into visible chromosomes, which is followed by the separation of the chromosomes into two identical sets. Cytokinesis can be considered as the last phase of mitosis when the two daughter cells separate, each with a nucleus and cytoplasmic organelles. Mitosis begins with nuclear membrane breakdown followed by condensation of the chromosomes and separation of the centrosomes (prophase). This is accompanied by the formation of mitotic spindles, which are attached on one end, the centrosomes, and at the other end, the kinetochore, a protein structure located at or near the centromeres of mitotic chromosomes (prometaphase). At this point, kinetochores that are unattached to the mitotic spindles generate a "wait" signal that delays the onset of anaphase until all chromosomes are properly attached and aligned. This signal is also called mitotic checkpoint or spindle assembly checkpoint. Following prometaphase, chromosomes congregate at the equatorial plate (metaphase) before separating to the opposite poles (anaphase). This is followed by the formation of new nuclear membranes around the daughter nuclei and uncoiling of the chromosomes (telophase), and finally into the formation of a cleavage furrow that leads to the formation of two daughter cells (cytokinesis).

PROPHASE

Prophase is the first phase of mitosis, the process that separates the duplicated genetic material carried in the nucleus of a parent cell into two identical daughter cells. During prophase, the complex of DNA and proteins contained in the nucleus, known as chromatin, condenses. The chromatin coils and becomes increasingly compact, resulting in the formation of visible chromosomes. Chromosomes are made of a single piece of DNA that is highly organized. The replicated chromosomes have an X shape and are called sister chromatids. The sister chromatids are pairs of identical copies of DNA joined at a point called the centromere. Then, a structure called the mitotic spindle begins to form. The mitotic spindle is made of long proteins called microtubules that begin forming at opposite ends of the cell. The spindle will be responsible for separating the sister chromatids into two cells. Prophase is followed by the second phase of mitosis, known as prometaphase.

The main events of prophase are: the condensation of chromosomes, the movement of the centrosomes, the formation of the mitotic spindle, and the beginning of nucleoli break down.

Condensation of Chromosomes

DNA that was replicated in interphase is condensed from molecules with lengths reaching 4 cm to chromosomes that are measured in micrograms. This process employs the condensin complex. Condensed chromosomes consist of two sister chromatids joined at the centromere.

Movement of Centrosomes

During prophase in animal cells, centrosomes move far enough apart to be resolved using a light microscope. Microtubule activity in each centrosome is increased due to recruitment of γ-tubulin. Replicated centrosomes from interphase move apart towards opposite poles of the cell, powered by centrosome associated motor proteins. Interdigitated interpolar microtubules from each centrosome interact with each other, helping to move the centrosomes to opposite poles.

Formation of the Mitotic Spindle

Microtubules involved in the interphase scaffolding break down as the replicated centrosomes separate. The movement of centrosomes to opposite poles is accompanied in animal cells by the organization of individual radial microtubule arrays (asters) by each centromere. Interpolar microtubules from both centrosomes interact, joining the sets of microtubules and forming the basic structure of the mitotic spindle. In cells without centrioles chromosomes can nucleate microtubule assembly into the mitotic apparatus. In plant cells, microtubules gather at opposite poles and begin to form the spindle apparatus at locations called foci. The mitotic spindle is of great importance in the process of mitosis and will eventually segregate the sister chromatids in metaphase.

Beginning of Nucleoli Breakdown

The nucleoli begin to break down in prophase, resulting in the discontinuation of ribosome production. This indicates a redirection of cellular energy from general cellular metabolism to cellular division. The nuclear envelope stays intact during this process.

PROMETAPHASE

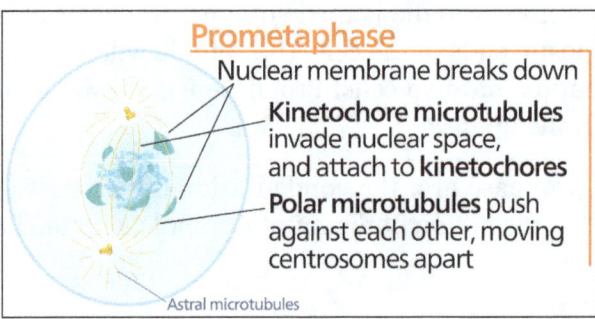

Prometaphase is the phase of mitosis following prophase and preceding metaphase, in eukaryotic somatic cells. In prometaphase, the nuclear membrane breaks apart into numerous "membrane vesicles", and the chromosomes inside form protein structures called kinetochores. Kinetochore microtubules emerging from the centrosomes at the poles (ends) of the spindle reach the chromosomes and attach to the kinetochores, throwing the chromosomes into agitated motion. Other spindle microtubules make contact with microtubules coming from the opposite pole. Forces exerted by protein "motors" associated with spindle microtubules move the chromosomes toward the centre of the cell.

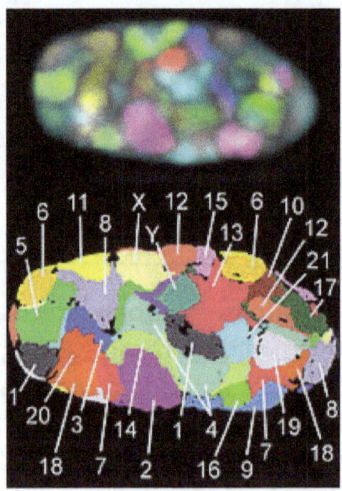

The 23 human chromosome territories during prometaphase in fibroblast cells.

Prometaphase is not always presented as a distinct part of mitosis. the events described here are instead assigned to late prophase and early metaphase.

Types of Microtubules

The microtubules are composed of two types, kinetochore microtubules and non-kinetochore microtubules:

- Kinetochore microtubules begin searching for kinetochores to attach to.

- A number of non-kinetochore microtubules or polar microtubules find and interact with corresponding nonkinetochore microtubules from the opposite centrosome to form the mitotic spindle.

Transition from Prometaphase to Metaphase

The role of prometaphase is completed when all of the kinetochore microtubules have attached to their kinetochores, upon which metaphase begins. An unattached kinetochore, and thus a non-aligned chromosome, even when most of the other chromosomes have lined up, will trigger the spindle checkpoint signal. This prevents premature progression into anaphase by inhibiting the anaphase-promoting complex until all kinetochores are attached and all the chromosomes aligned.

Early events of metaphase can coincide with the later events of prometaphase, as chromosomes with connected kinetochores will start the events of metaphase individually before other chromosomes with unconnected kinetochores that are still lingering in the events of prometaphase.

METAPHASE

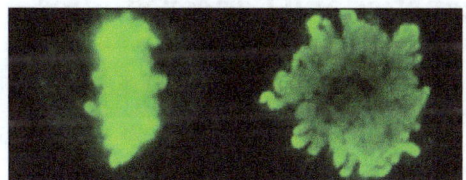

Chromosomes lined up on the metaphase plate. Two views with the metaphase plate rotated 60°.

Metaphase is a stage of mitosis in the eukaryotic cell cycle in which chromosomes are at their second-most condensed and coiled stage (they are at their most condensed in anaphase). These chromosomes, carrying genetic information, align in the equator of the cell before being separated into each of the two daughter cells. Metaphase accounts for approximately 4% of the cell cycle's duration. Preceded by events in prometaphase and followed by anaphase, microtubules formed in prophase have already found and attached themselves to kinetochores in metaphase.

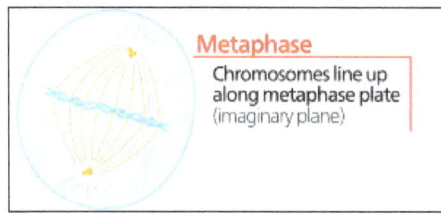

Metaphase in cells (here an animal cell) are characterized by the arrangement of chromosomes at the equatorial plane of the cell.

In metaphase, the centromeres of the chromosomes convene themselves on the metaphase plate (or equatorial plate), an imaginary line that is equidistant from the two centrosome poles. This even alignment is due to the counterbalance of the pulling powers generated by the opposing kinetochore microtubules, analogous to a tug-of-war between two people of equal strength, ending with the destruction of B cyclin. In certain types of cells, chromosomes do not line up at the metaphase plate and instead move back and forth between the poles randomly, only roughly lining up along the middleline. Early events of metaphase can coincide with the later events of prometaphase, as chromosomes with connected kinetochores will start the events of metaphase individually before other chromosomes with unconnected kinetochores that are still lingering in the events of prometaphase.

One of the cell cycle checkpoints occurs during prometaphase and metaphase. Only after all chromosomes have become aligned at the metaphase plate, when every kinetochore is properly attached to a bundle of microtubules, does the cell enter anaphase. It is thought that unattached or improperly attached kinetochores generate a signal to prevent premature progression to anaphase, even if most of kinetochores have been attached and most of the chromosomes have been aligned. Such a signal creates the mitotic spindle checkpoint. This would be accomplished by regulation of the anaphase-promoting complex, securin, and separase.

Metaphase in Cytogenetics and Cancer Studies

Human metaphase chromosomes (normal male karyotype).

The analysis of metaphase chromosomes is one of the main tools of classical cytogenetics and cancer studies. Chromosomes are condensed (thickened) and highly coiled in metaphase, which makes them most suitable for visual analysis. Metaphase chromosomes make the classical picture of chromosomes (karyotype). For classical cytogenetic analyses, cells are grown in short term culture and arrested in metaphase using mitotic inhibitor. Further they are used for slide preparation and banding (staining) of chromosomes to be visualised under microscope to study structure and number of chromosomes (karyotype). Staining of the slides, often with Giemsa (G banding) or Quinacrine, produces a pattern of in total up to several hundred bands. Normal metaphase spreads are used in methods like FISH and as a hybridization matrix for comparative genomic hybridization (CGH) experiments.

Malignant cells from solid tumors or leukemia samples can also be used for cytogenetic analysis to generate metaphase preparations. Inspection of the stained metaphase chromosomes allows the determination of numerical and structural changes in the tumor cell genome, for example, losses of chromosomal segments or translocations, which may lead to chimeric oncogenes, such as bcr-abl in chronic myelogenous leukemia.

ANAPHASE

Anaphase is the stage of mitosis after the process of metaphase, when replicated chromosomes are split and the newly-copied chromosomes (daughter chromatids) are moved to opposite poles of the cell. Chromosomes also reach their overall maximum condensation in late anaphase, to help chromosome segregation and the re-formation of the nucleus.

Anaphase starts when the anaphase promoting complex marks an inhibitory chaperone called securin for destruction by ubiquinylating it. Securin is a protein which inhibits a protease known as separase. The destruction of securin unleashes separase which then breaks down cohesin, a protein responsible for holding sister chromatids together. At this point, three subclasses of microtubule unique to mitosis are involved in creating the forces necessary to separate the chromatids: kinetochore microtubules, interpolar microtubules, and astral microtubules.

The centromeres are split, and the sister chromatids are pulled toward the poles by kinetochore microtubules. They take on a V-shape or Y-shape as they are pulled to either pole. While the chromosomes are drawn to each side of the cell, interpolar microtubules and astral microtubules generate forces that stretch the cell into an oval. Once anaphase is complete, the cell moves into telophase.

Phases

Anaphase is characterized by two distinct motions. The first of these, anaphase A, moves chromosomes to either pole of a dividing cell (marked by centrosomes, from which mitotic microtubules are generated and organised). The movement for this is primarily generated by the action of kinetochores, and a subclass of microtubule called kinetochore microtubules.

The second motion, anaphase B, involves the separation of these poles from each other. The movement for this is primarily generated by the action of interpolar microtubules and astral microtubules.

Anaphase A

A combination of different forces have been observed acting on chromatids in anaphase

A, but the primary force is exerted centrally. Microtubules attach to the midpoint of chromosomes (the centromere) via protein complexes (kinetochores). The attached microtubules depolymerise and shorten, which together with motor proteins creates movement that pulls chromosomes towards centrosomes located at each pole of the cell.

Anaphase B

The second part of anaphase is driven by its own distinct mechanisms. Force is generated by several actions. Interpolar microtubules begin at each centrosome and join at the equator of the dividing cell. They push against one another, causing each centrosome to move further apart. Meanwhile, astral microtubules begin at each centrosome and join with the cell membrane. This allows them to pull each centrosome closer to the cell membrane. Movement created by these microtubules is generated by a combination of microtubule growth or shrinking, and by motor proteins such as dyneins or kinesins.

Relation to the Cell Cycle

Anaphase accounts for approximately 1% of the cell cycle's duration. It begins with the regulated triggering of the metaphase-to-anaphase transition. Metaphase ends with the destruction of B cyclin. B cyclin is marked with ubiquitin which flags it for destruction by proteasomes, which is required for the function of metaphase cyclin-dependent kinases (M-Cdks). In essence, Activation of the Anaphase Promoting Complex (APC) causes the APC to cleave the M-phase cyclin and the inhibitory protein securin which activates the seperase protease to cleave the cohesin subunits holding the chromatids together.

TELOPHASE

Telophase is the final stage in both meiosis and mitosis in a eukaryotic cell. During telophase, the effects of prophase and prometaphase (the nucleolus and nuclear membrane disintegrating) are reversed. As chromosomes reach the cell poles, a nuclear envelope is re-assembled around each set of chromatids, the nucleoli reappear, and chromosomes begin to decondense back into the expanded chromatin that is present during interphase. The mitotic spindle is disassembled and remaining spindle microtubules are depolymerized. Telophase accounts for approximately 2% of the cell cycle's duration. Cytokinesis typically begins before late telophase and, when complete, segregates the two daughter nuclei between a pair of separate daughter cells.

Telophase is primarily driven by the dephosphorylation of mitotic cyclin-dependent kinase (Cdk) substrates.

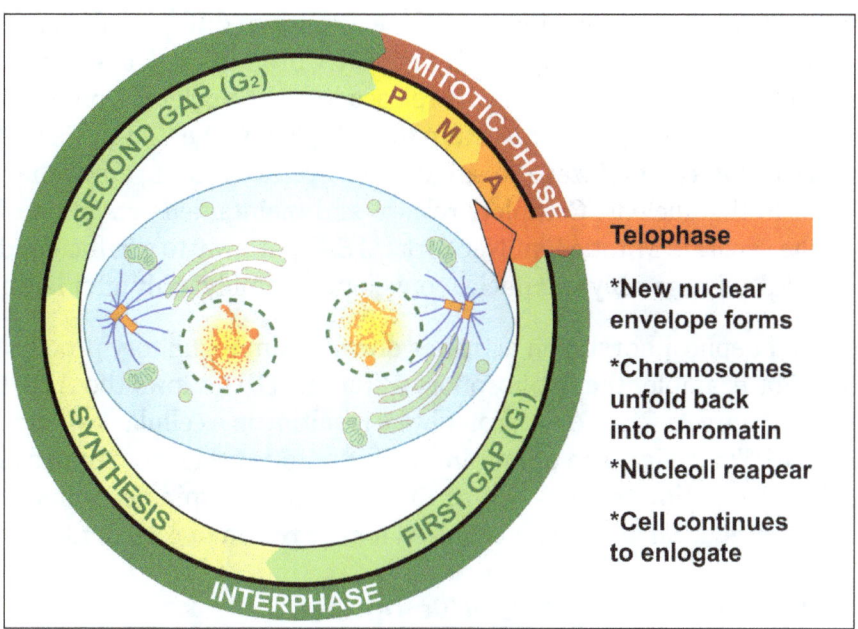

This image describes the final stage in mitosis, telophase.

Dephosphorylation of Cdk Substrates

The phosphorylation of the protein targets of M-Cdks (Mitotic Cyclin-dependent Kinases) drives spindle assembly, chromosome condensation and nuclear envelope breakdown in early mitosis. The dephosphorylation of these same substrates drives spindle disassembly, chromosome decondensation and the reformation of daughter nuclei in telophase. Establishing a degree of dephosphorylation permissive to telophase events requires both the inactivation of Cdks and the activation of phosphatases.

Cdk inactivation is primarily the result of the destruction of its associated cyclin. Cyclins are targeted for proteolytic degradation by the anaphase promoting complex (APC), also known as the cyclosome, a ubiquitin-ligase. The active, CDC20-bound APC (APC/C^{CDC20}) targets mitotic cyclins for degradation starting in anaphase. Experimental addition of non-degradable M-cyclin to cells induces cell cycle arrest in a post-anaphase/pre-telophase-like state with condensed chromosomes segregated to cell poles, an intact mitotic spindle, and no reformation of the nuclear envelope. This has been shown in frog (Xenopus) eggs, fruit flies (Drosophilla melanogaster), budding (Saccharomyces cerevisiae) and fission (Schizosaccharomyces pombe) yeast, and in multiple human cell lines.

The requirement for phosphatase activation can be seen in budding yeast, which do not have redundant phosphatases for mitotic exit and rely on the phosphatase cdc14. Blocking cdc14 activation in these cells results in the same phenotypic arrest as does blocking M-cyclin degradation.

Historically, it has been thought that anaphase and telophase are events that occur passively after satisfaction of the spindle-assembly checkpoint (SAC) that defines the metaphase-anaphase transition. However, the existence of differential phases to cdc14 activity between anaphase and telophase is suggestive of additional, unexplored late-mitotic checkpoints. Cdc14 is activated by its release into the nucleus, from sequestration in the nucleolus, and subsequent export into the cytoplasm. The Cdc-14 Early Anaphase Release pathway, which stabilizes the spindle, also releases cdc14 from the nucleolus but restricts it to the nucleus. Complete release and maintained activation of cdc14 is achieved by the separate Mitotic Exit Network (MEN) pathway to a sufficient degree (to trigger the spindle disassembly and nuclear envelope assembly) only after late anaphase.

Cdc14-mediated dephosphorylation activates downstream regulatory processes unique to telophase. For example, the dephosphorylation of CDH1 allows the APC/C to bind CDH1. APC/C^{CDH1} targets CDC20 for proteolysis, resulting in a cellular switch from APC/C^{CDC20} to APC/C^{CDH1} activity. The ubiquitination of mitotic cyclins continues along with that of APC/C^{CDH1}-specific targets such as the yeast mitotic spindle component, Ase1, and cdc5, the degradation of which is required for the return of cells to the G1 phase.

Additional Mechanisms Driving Telophase

A shift in the whole-cell phosphoprotein profile is only the broadest of many regulatory mechanisms contributing to the onset of individual telophase events:

- The anaphase-mediated distancing of chromosomes from the metaphase plate may trigger spatial cues for the onset of telophase.

- An important regulator and effector of telophase is cdc48 (homologous to yeast cdc48 is human p97, both structurally and functionally), a protein that mechanically employs its ATPase activity to alter target protein conformation. Cdc48 is necessary for spindle disassembly, nuclear envelope assembly, and chromosome decondensation. Cdc48 modifies proteins structurally involved in these processes and also some ubiquitinated proteins which are thus targeted to the proteasome.

Mitotic Spindle Disassembly

The breaking of the mitotic spindle, common to the completion of mitosis in all eukaryotes, is the event most often used to define the anaphase-B to telophase transition, although the initiation of nuclear reassembly tends to precede that of spindle disassembly.

Spindle disassembly is an irreversible process which must effect not the ultimate degradation, but the reorganization of constituent microtubules; microtubules are detached from kinetochores and spindle pole bodies and return to their interphase states.

Spindle depolymerization during telophase occurs from the plus end and is, in this way, a reversal of spindle assembly. Subsequent microtubule array assembly is, unlike that of the polarized spindle, interpolar. This is especially apparent in animal cells which must immediately, following mitotic spindle disassembly, establish the antiparallel bundle of microtubules known as the central spindle in order to regulate cytokinesis. The ATPase p97 is required for the establishment of the relatively stable and long interphase microtubule arrays following disassembly of the highly dynamic and relatively short mitotic ones.

While spindle assembly has been well studied and characterized as a process where tentative structures are edified by the SAC, the molecular basis of spindle disassembly is not understood in comparable detail. The late-mitotic dephosphorylation cascade of M-Cdk substrates by the MEN is broadly held to be responsible for spindle disassembly. The phosphorylation states of microtubule stabilizing and destabilizing factors, as well as microtubule nucleators are key regulators of their activities. For example, NuMA is a minus-end crosslinking protein and Cdk substrate whose dissociation from the microtubule is effected by its dephosphorylation during telophase.

A general model for spindle disassembly in yeast is that the three functionally overlapping subprocesses of spindle disengagement, destabilization, and depolymerization are primarily effected by APC/C^{CDH1}, microtubule-stabilizer-specific kinases, and plus-end directed microtubule depolymerases, respectively. These effectors are known to be highly conserved between yeast and higher eukaryotes. The APC/C^{CDH1} targets crosslinking microtubule-associated proteins (NuMA, Ase1, Cin1 and more). AuroraB (yeast Ipl1) phosphorylates the spindle-associated stabilizing protein EB1 (yeast Bim1), which then dissociates from microtubules, and the destabilizer She1, which then associates with microtubules. Kinesin8 (yeast Kip3), an ATP-dependent depolymerase, accelerate microtubule depolymerization at the plus end. It was shown the concurrent disruption of these mechanisms, but not of any one, results in dramatic spindle hyperstability during telophase, suggesting functional overlap despite the diversity of the mechanisms.

Nuclear Envelope Reassembly

The main components of the nuclear envelope are a double membrane, nuclear pore complexes, and a nuclear lamina internal to the inner nuclear membrane. These components are dismantled during prophase and prometaphase and reconstructed during telophase, when the nuclear envelope reforms on the surface of separated sister chromatids. The nuclear membrane is fragmented and partly absorbed by the endoplasmic reticulum during prometaphase and the targeting of inner nuclear membrane protein-containing ER vesicles to the chromatin occurs during telophase in a reversal of this process. Membrane-forming vesicles aggregate directly to the surface of chromatin, where they fuse laterally into a continuous membrane.

Ran-GTP is required for early nuclear envelope assembly at the surface of the chromosomes: It releases envelope components sequestered by importin β during early mitosis. Ran-GTP localizes near chromosomes throughout mitosis, but does not trigger the dissociation of nuclear envelope proteins from importin β until M-Cdk targets are dephosphorylated in telophase. These envelope components include several nuclear pore components, the most studied of which is the nuclear pore scaffold protein ELYS, which can recognize DNA regions rich in A:T base pairs (in vitro), and may therefore bind directly to the DNA. However, experiments in Xenopus egg extracts have concluded that ELYS fails to associate with bare DNA and will only directly bind histone dimers and nucleosomes. After binding to chromatin, ELYS recruits other components of the nuclear pore scaffold and nuclear pore trans-membrane proteins. The nuclear pore complex is assembled and integrated in the nuclear envelope in an organized manner, consecutively adding Nup107-160, POM121, and FG Nups.

It is debated whether the mechanism of nuclear membrane reassembly involves initial nuclear pore assembly and subsequent recruitment of membrane vesicles around the pores or if the nuclear envelope forms primarily from extended ER cisternae, preceding nuclear pore assembly:

- In cells where the nuclear membrane fragments into non-ER vesicles during mitosis, a Ran-GTP–dependent pathway can direct these discrete vesicle populations to chromatin where they fuse to reform the nuclear envelope.

- In cells where the nuclear membrane is absorbed into the endoplasmic reticulum during mitosis, reassembly involves the lateral expansion around the chromatin with stabilization of the expanding membrane over the surface of the chromatin. Studies claiming this mechanism is a prerequisite to nuclear pore formation have found that bare-chromatin-associated Nup107–160 complexes are present in single units instead of as assembled pre-pores.

The envelope smoothens and expands following its enclosure of the whole chromatid set. This probably occurs occurs due to the nuclear pores' import of lamin, which can be retained within a continuous membrane. The nuclear envelopes of Xenopus egg extracts failed to smoothen when nuclear import of lamin was inhibited, remaining wrinkled and closely bound to condensed chromosomes. However, in the case of ER lateral expansion, nuclear import is initiated before completion of the nuclear envelope reassembly, leading to a temporary intra-nuclear protein gradient between the distal and medial aspects of the forming nucleus.

Lamin subunits disassembled in prophase are inactivated and sequestered during mitosis. Lamina reassembly is triggered by lamin dephosphorylation (and additionally by methyl-esterification of COOH residues on lamin-B). Lamin-B can target chromatin as early as mid-anaphase. During telophase, when nuclear import is reestablished, lamin-A enters the reforming nucleus but continues to slowly assemble into the peripheral lamina over several hours in throughout the G1 phase.

Xenopus egg extracts and human cancer cell lines have been the primary models used for studying nuclear envelope reassembly. Yeast lack lamins; their nuclear envelope remains intact throughout mitosis and nuclear division happens during cytokinesis.

Chromosome Decondensation

Chromosome decondensation (also known as relaxation or decompaction) into expanded chromatin is necessary for the cell's resumption of interphase processes, and occurs in parallel to nuclear envelope assembly during telophase in many eukaryotes. MEN-mediated Cdk dephosphorylation is necessary for chromosome decondensation.

In vertebrates, chromosome decondensation is initiated only after nuclear import is reestablished. If lamin transport through nuclear pores is prevented, chromosomes remain condensed following cytokinesis, and cells fail to reenter the next S phase. In mammals, DNA licensing for S phase (the association of chromatin to the multiple protein factors necessary for its replication) also occurs coincidentally with the maturation of the nuclear envelope during late telophase. This can be attributed to and provides evidence for the nuclear import machinery's reestablishment of interphase nuclear and cytoplasmic protein localizations during telophase.

FACTORS THAT AFFECT MITOSIS

Growth and Repair

After an injury many cells are replaced in order to repair the damage. The rate of mitosis must increase in order to produce these new cells. Similarly the rate of mitosis also increases during periods of growth, such as our development in the womb, childhood and puberty. In pants there is also much higher rates of mitosis during germination and growth.

Nutrient Availability

Nutrients are needed as a source of energy and as building blocks. Cell will need DNA nucleotides in order to synthesise new DNA and the cell organelles will have to be copied too. A lack of necessary nutrients can restrict or stop mitosis. This is why people in areas of famine are often smaller and shorter.

Cell Type and Location

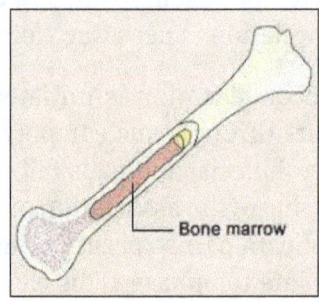

Bone marrow

Body tissues that are replaced frequently have a higher rate of mitosis. For example bone marrow which is responsible for blood cell production is an area in the body where rapid mitosis occurs. Skin cells, hair follicles and the cells lining our intestines (epithelial cells) all have high rates of mitosis as these tissues constantly need to be replaced. In plants growth occurs largely at the shoot and root tips. These cells have much higher rates of mitosis than the rest of the plant.

Damaged tissues are not simply replaced by surrounding cells. Special stem cells do this. Stem cells are undifferentiated cells, this means that they have not fully developed into a specific cell type and can divide to produce different cell types. During growth or repair stem cells undergo mitosis and cell division to produce more of whatever cell type is required. Stem cells have special mechanisms that allows them to divide many more times than regular cells.

Enzyme Activity

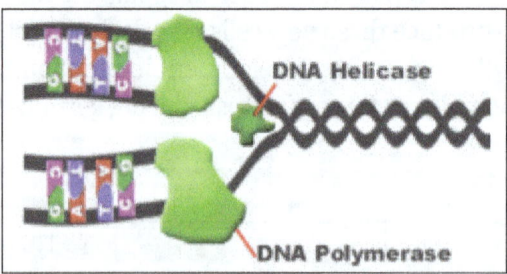

DNA Helicase

DNA Polymerase

DNA replication, mitosis and cell division are controlled by an array of proteins and enzymes. Therefore the factors that affect enzymes can also affect mitosis. Some of the key factors that affect enzyme activity include:

- Temperature.

- pH.

- Substrate concentration (e.g. the availability of nutrient building blocks).

- Enzyme Cofactors (some dietary vitamins are essential for enzyme function).

Regulation

The cell cycle, including growth, DNA replication, mitosis and cell division are all highly regulated processes that are regulated by special genes within a cell. Mutations in these genes can result in a loss of regulation and may cause a larger number of these cells to grow. This causes a lump known as a tumour

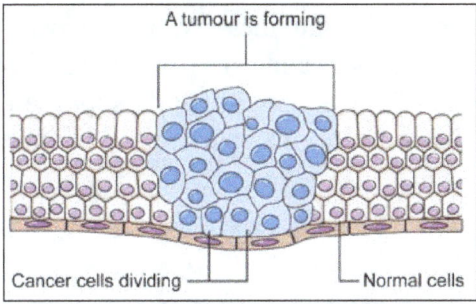

Patients undergoing chemotherapy often lose their hair, feel nauseous and may need a bone-marrow transplant. This is because chemotherapy drugs target cells with higher mitosis/division rates such as tumour cells. However, the side effect is that it also affects body cells that normally have a higher rate of mitosis/division.

References

- Singh, Ram J. (2017). Plant Cytogenetics, Third Edition. Boca Raton, FL: CBC Press, Taylor & Francis Group. p. 19. ISBN 9781439884188

- Mitosis, science-243070: britannica.com, Retrieved 10 January, 2020

- Asbury CL (February 2017). "Anaphase A: Disassembling Microtubules Move Chromosomes toward Spindle Poles". Biology. 6 (1): 15. doi:10.3390/biology6010015. PMC 5372008. PMID 28218660

- Factors-that-affect-mitosis, Plain, Tree: pathwayz.org, Retrieved 30 April, 2020

- May, Karen M; Kevin G. Hardwick (2006). "The spindle checkpoint". Journal of Cell Science. 119 (Pt 20): 4139–42. doi:10.1242/jcs.03165. PMID 17038540. Retrieved 9 December 2012

4

Phases in Meiosis

The phases of meiosis can be divided into meiosis I and meiosis II which are further divided into karyokinesis I and cytokinesis I and karyokinesis II and cytokinesis II respectively. It involves prophase 1, metaphase 1, anaphase 1 and telophase 1. The topics elaborated in this chapter will help in gaining a better perspective of the different phases of meiosis.

MEIOSIS 1

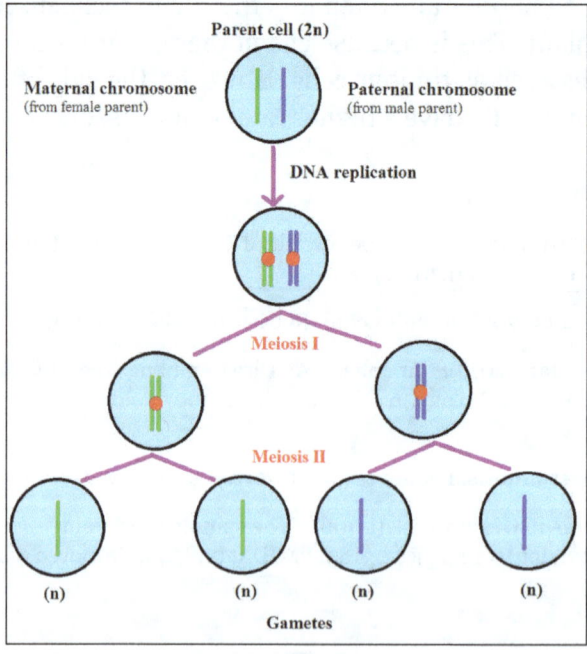

Meiosis 1 Stages.

Mitotic cell division is equational in nature while meiosis is a reduction division. The salient features of meiotic division that make it different from mitosis are as follows:

- It occurs in two stages of the nuclear and cellular division as Meiosis I and Meiosis II. DNA replication occurs, however, only once.

- It involves the pairing of homologous chromosomes and recombination between them.

- Four haploid daughter cells are produced at the end, unlike two diploid daughter cells in mitosis.

Meiosis 1 separates the pair of homologous chromosomes and reduces the diploid cell to haploid. It is divided into several stages that include, prophase, metaphase, anaphase and telophase.

The different stages of meiosis 1 can be explained by the following phases :

- Prophase 1.

- Metaphase 1.

- Anaphase 1.

- Telophase.

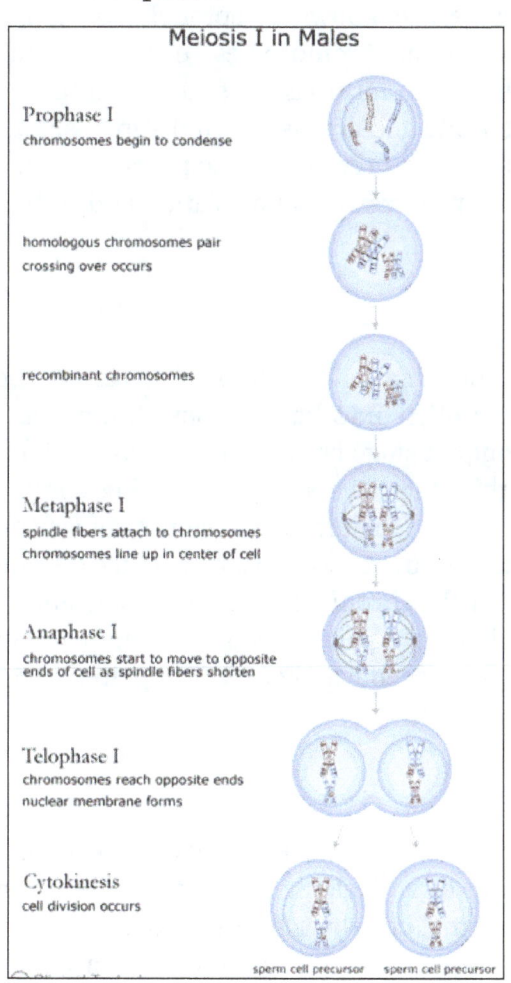

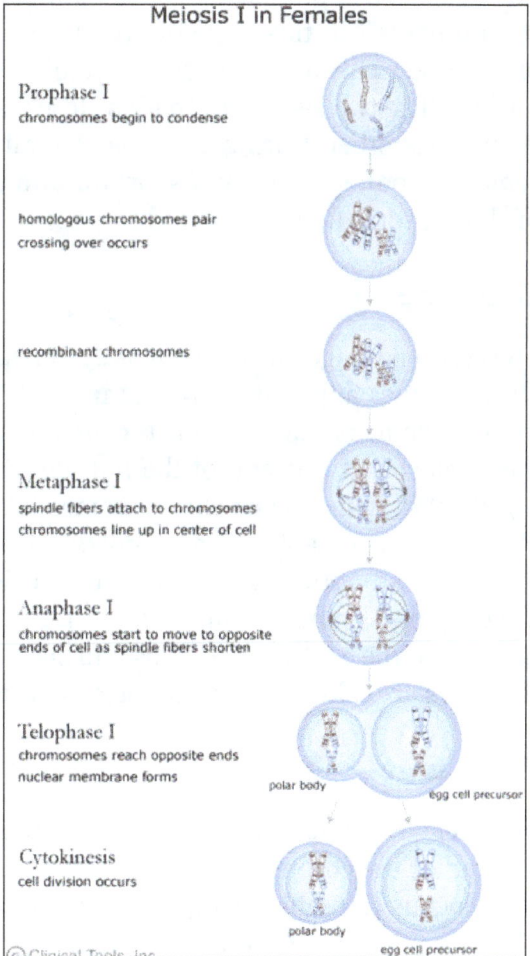

Prophase I

Prophase I is typically the longest phase of meiosis. During prophase I, homologous chromosomes pair and exchange genetic information (homologous recombination). This often results in chromosomal crossover. This process facilitates pairing between homologous chromosomes and hence accurate segregation of the chromosomes at the first meiosis division. The new combinations of DNA created during crossover are a significant source of genetic variation, and result in new combinations of alleles, which may be beneficial. The paired and replicated chromosomes are called bivalents or tetrads, which have two chromosomes and four chromatids, with one chromosome coming from each parent. The process of pairing the homologous chromosomes is called synapsis. At this stage, non-sister chromatids may cross-over at points called chiasmata (plural; singular chiasma). Prophase I has historically been divided into a series of substages which are named according to the appearance of chromosomes.

Leptotene

The first stage of prophase I is the leptotene stage, also known as leptonema, meaning "thin threads". In this stage of prophase I, individual chromosomes—each consisting of two sister chromatids—become "individualized" to form visible strands within the nucleus. The two sister chromatids closely associate and are visually indistinguishable from one another. During leptotene, lateral elements of the synaptonemal complex assemble. Leptotene is of very short duration and progressive condensation and coiling of chromosome fibers takes place.

Zygotene

The zygotene stage, also known as zygonema, meaning "paired threads", occurs as the chromosomes approximately line up with each other into homologous chromosome pairs. In some organisms, this is called the bouquet stage because of the way the telomeres cluster at one end of the nucleus. At this stage, the synapsis (pairing/coming together) of homologous chromosomes takes place, facilitated by assembly of central element of the synaptonemal complex. Pairing is brought about in a zipper-like fashion and may start at the centromere (procentric), at the chromosome ends (proterminal), or at any other portion (intermediate). Individuals of a pair are equal in length and in position of the centromere. Thus pairing is highly specific and exact. The paired chromosomes are called bivalent or tetrad chromosomes.

Pachytene

The pachytene stage, also known as pachynema, meaning "thick threads". At this point a tetrad of the chromosomes has formed known as a bivalent. This is the stage when homologous recombination, including chromosomal crossover (crossing over), occurs. Nonsister chromatids of homologous chromosomes may exchange genetic information over regions of homology. DNA damage induced by gamma radiation during the lepto-tene to early pachytene stages induces an homologous recombinational repair (HRR) pathway that employs the key proteins DMC1 and RAD51. This HRR pathway is replaced at mid-pachytene by the less accurate repair pathway of non-homologous end joining and an HRR pathway that does not depend on DMC1. Sex chromosomes, however, are not wholly identical, and only exchange information over a small region of homology. At the sites where exchange happens, chiasmata form. The exchange of information between the non-sister chromatids results in a recombination of information; each chromosome has the complete set of information it had before, and there are no gaps formed as a result of the process. Because the chromosomes cannot be distinguished in the synaptonemal complex, the actual act of crossing over is not perceivable through the microscope, and chiasmata are not visible until the next stage.

Diplotene

During the diplotene stage, also known as diplonema, meaning "two threads", the syn-aptonemal complex degrades and homologous chromosomes separate from one anoth-er a little. The chromosomes themselves uncoil a bit, allowing some transcription of DNA. However, the homologous chromosomes of each bivalent remain tightly bound at chiasmata, the regions where crossing-over occurred. The chiasmata remain on the chromosomes until they are severed at the transition to anaphase I.

In human fetal oogenesis, all developing oocytes develop to this stage and are arrested in prophase I before birth. This suspended state is referred to as the dictyotene stage or dictyate. It lasts until meiosis is resumed to prepare the oocyte for ovulation, which happens at puberty or even later.

Diakinesis

Chromosomes condense further during the diakinesis stage, meaning "moving through". This is the first point in meiosis where the four parts of the tetrads are ac-tually visible. Sites of crossing over entangle together, effectively overlapping, making chiasmata clearly visible. Other than this observation, the rest of the stage closely re-sembles prometaphase of mitosis; the nucleoli disappear, the nuclear membrane disin-tegrates into vesicles, and the meiotic spindle begins to form.

Synchronous Processes

During these stages, two centrosomes, containing a pair of centrioles in animal cells, migrate to the two poles of the cell. These centrosomes, which were duplicated during S-phase, function as microtubule organizing centers nucleating microtubules, which are essentially cellular ropes and poles. The microtubules invade the nuclear region after the nuclear envelope disintegrates, attaching to the chromosomes at the kinetochore. The ki-netochore functions as a motor, pulling the chromosome along the attached microtubule toward the originating centrosome, like a train on a track. There are four kinetochores on each tetrad, but the pair of kinetochores on each sister chromatid fuses and functions as a unit during meiosis I.

Microtubules that attach to the kinetochores are known as kinetochore microtubules. Other microtubules will interact with microtubules from the opposite centrosome: these are called nonkinetochore microtubules or polar microtubules. A third type of microtubules, the aster microtubules, radiates from the centrosome into the cytoplasm or contacts components of the membrane skeleton.

Metaphse I

The first metaphase of meisosis I encompasses the alignment of paired chromosomes along the center (metaphase plate) of a cell, ensuring that two complete copies of chromosomes are present in the resulting two daughter cells of meiosis I. Metaphase I follows prophase I and precedes anaphase I.

In meiosis I, the lining-up stage of metaphase I is relatively rapid. Tetrads or bivalents (a pair of chromosomes with four chromatids (2 originals, 2 copies) are pulled into line at what is known as the metaphase (or equatorial) plate. This plate does not actually exist but is an imaginary central line along which the chromosomes are positioned.

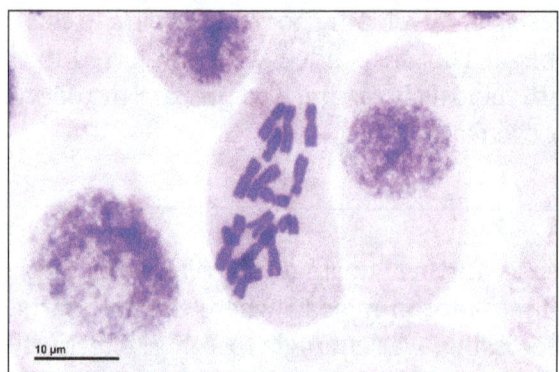

Metaphase in an onion cell.

The individual chromosome in each pair remains close to its partner and lines up one on top of the other. This will eventually lead to one chromosome migrating to one pole, the other to the opposite pole. It also does not matter in which direction these chromosomes are horizontally oriented along this imaginary line. The DNA from either parent can face either side of the cell. This increases gene variation, as where one daughter cell might contain 40% of the father's chromosomes and 60% of the mother's, the other will have 60% and 40% of each parent respectively. This would account, for example, for a first child having the father's eyes and the mother's nose, and the second having the mother's eyes and the father's nose. Due to recombination in the prophase, either child will not look exactly the same as either parent, only similar.

In the image below of an onion cell (in metaphase I), the dark-purple stained chromosome pairs are all centrally positioned along the metaphase plate. In metaphase I, the two chromosomes of a homologous pair face opposite poles. As recombination has taken place, each of the four chromatids (and, of course, both homologous pairs) have slightly different genetic material. Further steps will pull one of the homologous pair to one end of the cell, and the other to the opposite end. This means that the resulting two cells (produced during cytokinesis at the end of meiosis I) will contain full sets of chromosomes, but these will not consist of chromosome pairs.

Confusion Surrounding Chromosomes and Chromatids

This allocation of chromosomes is particularly important to understand but often leads to much confusion. The human karyotype is composed of 22 chromosome pairs and one pair of sex chromosomes (XX or XY). A total of 23 chromosome pairs. Every non-gamete cell contains this set of 23 pairs within its nucleus, except for mature red blood cells or old cornified cells. The two chromosomes that make up a homologous pair come from an organism's parents – one from the father and one from the mother. In the karyotype below of a male human, these pairs are very distinct. The presence of 23 pairs, and a total of 46 chromosomes.

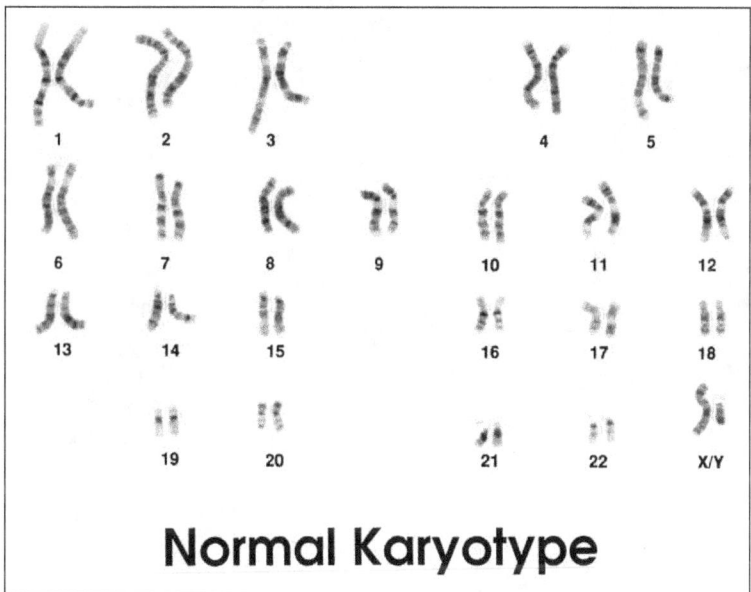

Human male chromosomes.

In this image, no X-shape can be seen. Each chromosome is a single strand. Each single strand is a chromatid with a centromere (not visible). This is because the X-shape exists only after the replication of DNA occurs during the S stage of the interphase which precedes both mitosis and meiosis.

The great confusion many students encounter when studying meiosis is the traditional idea that chromosomes are X-shaped. However, this is only the case after each chromatid has been replicated to create two sister chromatids joined together at the centromere. Before replication, a human set of 23 chromosome pairs (each pair consisting of one chromosome from the father, one from the mother) has 46 chromosomes, as we can count in the above image. Each of these 46 chromosomes consists of a single chromatid. The human karyotype before replication, therefore, has 46 chromosomes (23 pairs) and 46 chromatids.

After the S stage of interphase, the chromatids are replicated and the copies remain attached to the 'sister' via the centromere. The final count is still 23 chromosome pairs, but each chromosome now consists of two sister chromatids. This extra chromatid provides the X-shape of the chromosome as we are used to seeing it. There are still 46 chromosomes (23 pairs), but now there are 92 chromatids. As the combination of two chromosomes (a pair) containing 4 chromatids (two original, two replicated) is also known as either a tetrad or a bivalent, we can also say that a tetrad has two chromosomes and four chromatids. Tetrás means four, while bi relates to two.

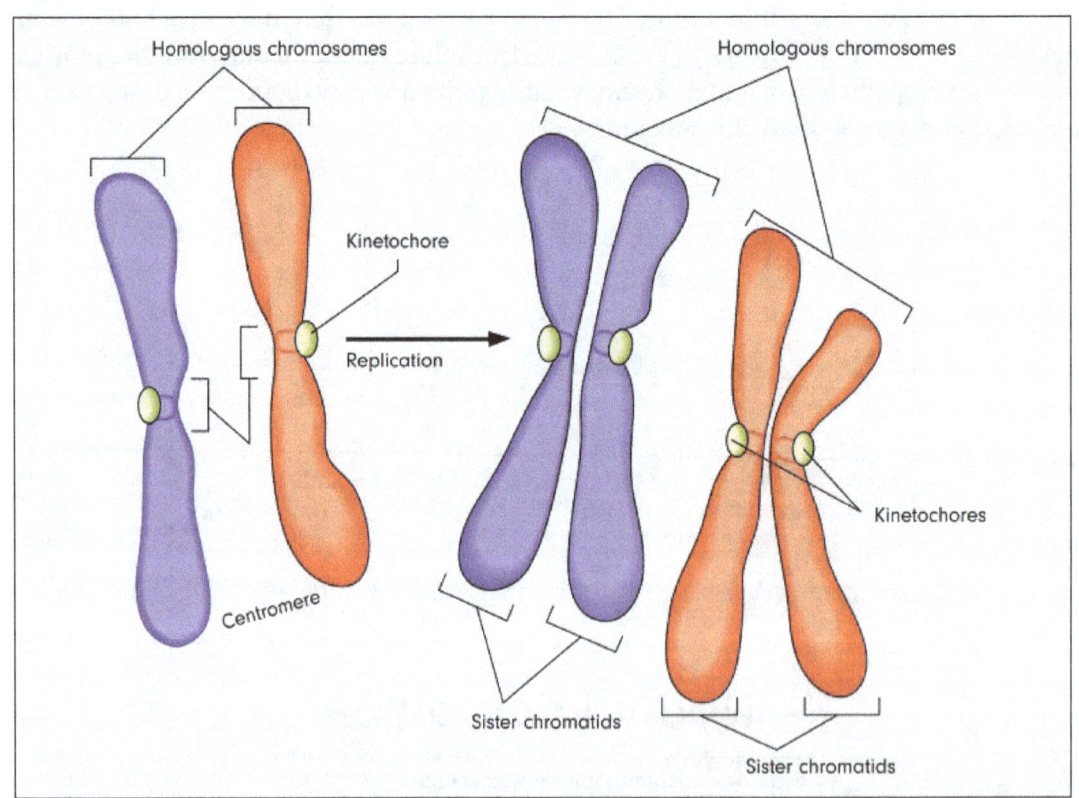

Chromosome pair before and after replication.

This all becomes much clearer when we think of the goal of meiosis and the subsequent fertilization of an egg by a sperm. Meiosis is a two-step process (with many sub-steps) that first splits a single cell into two cells, each with a complete but slightly different set of DNA contained in chromosome pairs. This is why DNA must be replicated beforehand. If not, it would not be possible to provide a full set of chromosome pairs to the two daughter cells. The end result is two daughter cells, each containing 46 chromosomes (23 pairs) and 46 chromatids.

These two cells then split a second time under meiosis II, creating four daughter cells which each contain a complete set of DNA which is not presented in chromosome pairs, but as a single chromosome. Each cell now contains 23 chromosomes and 23 chromatids. These cells are produced as sperm in males and eggs in females. Upon fertilization, the single chromosomes of each sex combine, once again creating a full chromosome pair.

Chromosome Alignment along the Metaphase Plate

The imaginary line through the center of a cell called the metaphase plate requires the spindle apparatus in order to align the matching chromosome pairs along its length.

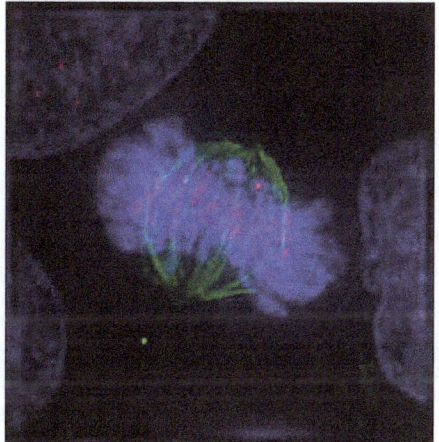

Green and blue-lit spindle structure.

Centrioles, protein structures that send out fibers known as microtubules, position themselves at opposite sides of the cell, creating two poles. Long protein fibers called microtubules are then sent out from both centrioles, forming the spindle apparatus. These microtubules meet (but do not join together) at the center of the cell where the chromosome pairs line up. Once the spindle apparatus is constructed, an attachment occurs between each chromosome's kinetochore, found close its centromere. This attachment is very stable and has been compared to a Chinese finger trap – the higher the tension, the stronger the attachment.

If we imagine the centrioles lie at the north and south of a circular cell, with the chromosome pairs lined up through the equivalent of the equator – one of the pair above the equator and the other below – it is easy to also imagine how these chromosome pairs can be separated, one pulled up, the other down along these fibers. This migration does not happen during the metaphase. Instead, microtubules pull at the chromosomes until they are properly aligned. The cell then checks the chromosomes are correctly positioned before moving on to the next phase of meiosis I.

Anaphase I

Anaphase I is the third stage of meiosis I and follows prophase I and metaphase I. This stage is characterized by the movement of chromosomes to both poles of a meiotic cell via a microtubule network known as the spindle apparatus. This mechanism separates homologous chromosomes into two separate groups.

In anaphase I, the main goal of the spindle apparatus is apparent. This phase can only take place after a positive spindle checkpoint result at the end of the preceding phase, the metaphase. This check make's sure that the kinetochores – the equivalent of climbing equipment that enables chromosomes to work their way along the microtubule network emitting from the centrioles – are properly attached. Kinetochores are proteins that connect microtubules to the chromosome centromere. Only once all chromosomes are tightly attached to the spindle network can anaphase I begin.

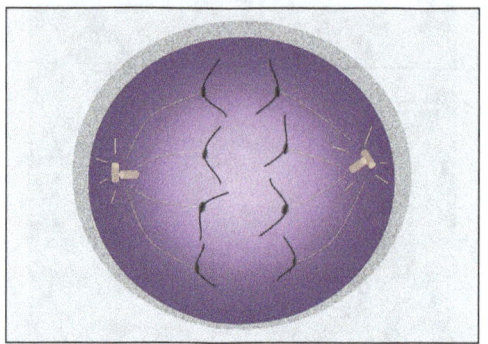

Anaphase.

Preparation for Anaphase I

Centrosomes have been commonly believed to be the primary manufacturing units of the spindle apparatus. Centrioles are the main components of the centrosome and appear as short microtubule cylinders in a star-like assembly consisting of groups of three microtubules, as seen in the image below. Centrioles exist in pairs. When combined with pericentriolar material (PCM) they form a centrosome.

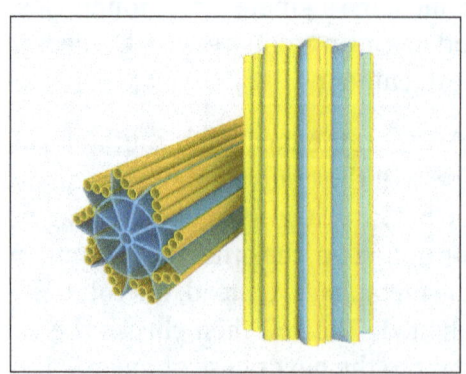

A Pair of Centrioles

A centriole pair is not attached, but during the replication stage (synthesis, or S-stage) of the cell cycle, they are also replicated to form daughter centrioles. The original and daughter centrioles are attached and will remain so until the prophase stage of mitosis or prophase I of meiosis is initiated. It is important that four structures are available, as these will split into two pairs – one pair for each daughter cell.

During prophase, centrosomes separate and migrate to two opposite poles outside of the disintegrating nuclear membrane, where they produce longer microtubules that reach to the cell center (the metaphase plate or equatorial plate). Microtubules have a 'slow-growing' (minus) end attached to the PCM of each centrosome, and a 'fast-growing' (plus) end which grows towards the metaphase plate.

During prometaphase, kinetochore proteins form close to the centromeres of the chromosomes and attach to the nearest microtubule. If a centromere is damaged, no kinetochore will form and there will be no anaphase. One of each chromosome pair is destined to be carried to one pole, the other to the opposite pole.

During metaphase, each chromosome is aligned and ready to be transported to either pole. As mentioned, a positive result to a spindle check must be completed before the move into anaphase, in which chromosome pairs are separated from their partners and transported to either side of the cell.

What happens during Anaphase I?

How the chromosomes are brought to each side of the cell during anaphase is now understood as a breakdown of the microtubule network, shortening the microtubule (MT) fibers and hence bringing each chromosome closer to its final destination. Kinetochore function is therefore crucial, as these keep the chromosomes attached, like abseiling rope clips, to a rope that may fray at any moment. As most kinetochores are attached to more than one microtubule – usually a bundle – the breakdown of single sections of microtubule does not lead to the detachment of a chromosome from the spindle, yet this attachment must still be strong. Some scientists liken the kinetochore to microtubule bundle connection to the Chinese finger trap, where any traction forces create an even stronger attachment.

During anaphase I the different microtubule types and functions become important. Kinetochore, astral and interpolar MTs are the three types of microtubule essential for spindle formation. They are formed from layers of alpha and beta tubulin proteins, as shown below.

Kinetochore MTs are centrosome fibers connected to chromosome centromeres via a kinetochore. Interpolar MTs are long microtubules that extend past the metaphase plate, overlapping the positive ends of interpolar MTs coming from the opposite pole. Astral MTs do not connect to kinetochores and are thought to create a navigation network for other areas of the spindle network.

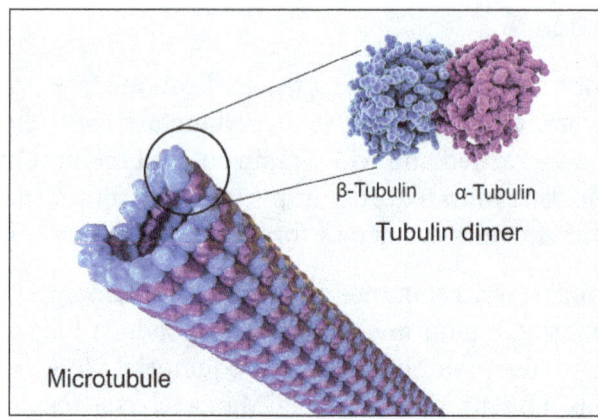

β-Tubulin   α-Tubulin

Tubulin dimer

Microtubule

Microtubule structure.

Anaphase actually consists of two stages: Anaphase A and B. These occur simultaneously but are very different mechanisms. In anaphase A, the connecting fibers of the microtubule spindle shorten through the breaking up of small sections, while kinetochores lead their chromosomes up- or downwards. Electron microscopy usually shows the kinetochore attachment point closer to the pole, with the chromosome arms dangling in the opposite direction. This shows that there are other forces at work in vertebrates which create 'drag' on the chromosome arms. These forces are called polar winds. Kinetochores must, therefore, be constantly in motion to prevent a reversal of direction or a to-ing and fro-ing along the spindle, otherwise known as directional instability. This is similar to dynamic instability during anaphase A, caused by the switch between breakage and growth of the microtubules – breakage (shortening) to bring chromosomes closer to their destination, growth (lengthening) to separate the poles even further.

In anaphase B, the spindle network elongates, separating the cell poles even further. Overlapping sections of microtubules – plus ends originating from both poles – slide apart. This mechanism is highly regulated. The cell stretches and elongates, whereby the poles become further apart. This helps to prevent incomplete cell division. Once a complete set of chromosomes has arrived at either pole of the cell, the next phase – telophase – may begin.

Spindle Apparatus Formation and Organization – Are Centrioles Necessary?

As the organization of the structures necessary for cell division are of great interest to fertility experts, much research has been done concerning spindle construction. An improperly attached chromosome or incompletely developed spindle network during meiosis leads to infertility or miscarriage. The dissolving of the nuclear membrane, allowing centrosomes to migrate to both cell poles, is not timed according to spindle development. This means that the spindle network might not be completed in time for the separation of the chromosome pairs in meiosis I, or of the single chromosomes in meiosis II. Alternatively, early disintegration of the nuclear membrane means that the spindle apparatus can be completely formed prior to anaphase.

Cells that normally contain centrosomes can form spindle networks even after their centrioles have been artificially removed in a laboratory setting. Many plant cells and all oocytes (egg cells) do not contain centrosomes, but microtubule networks still form inside them. It is suggested that the microtubule network is itself responsible for spindle formation, and not the centrosomes. The fact that centrosomes are also responsible for flagellum and cilia formation, and that all male vertebrae cells have centrosomes, might point towards the centrosome as being necessary for flagella, but only complementary for spindle formation. As the ova is a very large gamete, spindle formation from microtubules already within the cytoplasm might be more energy efficient.

However, the large size of the egg can create other problems. At least 10% of human pregnancies produce aneuploid embryos with either extra or missing chromosomes. The most commonly quoted of these is trisomy 21, or Down syndrome. This is because atypical spindle forming and positioning can lead to asymmetrical cell division and the loss or gain of one or more chromosomes to either daughter cell.

The image below shows the large difference in size between egg ova and spermatozoa, and the long flagellum of the male gamete. The presence of centrioles in the male gamete and not the female may therefore be more important for flagellum formation than the spindle network.

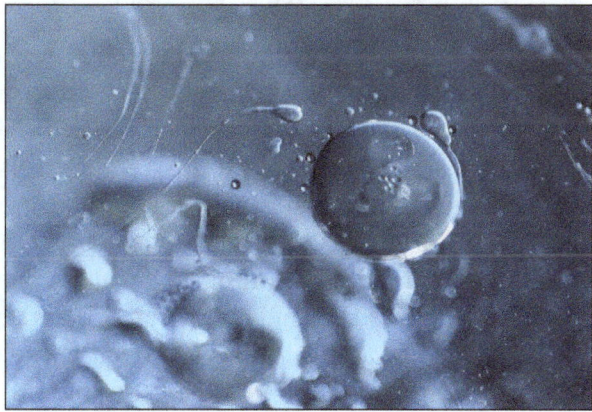

Sperm and egg – Electron microscopy.

Separating the Tetrad

Due to replication during a precursory phase in mitosis (non-gamete cell division) the chromosome pairs separated during anaphase I are different from the chromosome pairs in the normal human karyotype.

Before preparation for either mitosis (somatic cells) or meiosis (gametes or sex-cells) a human cell contains 46 loosely wrapped chromatids within its nucleus. These are arranged in pairs – one from the mother and one from the father. This means the non-dividing human cell contains 23 pairs of chromatids, the total sum of which is 46 chromatids. The sole difference between a chromatid and a chromosome is the packaging – when loosely packaged the complex of DNA and binding proteins is called a chromatid, when tightly packaged it is called a chromosome.

A tetrad (or bivalent) is specific to the process of meiosis. It is the result of replication in the S-phase of the natural cell cycle in combination with the meiosis-specific procedure of crossing over, the result of which is recombination – the mixing up of alleles across a chromosome pair. Upon recombination, a pair of chromosomes is referred to as a tetrad. In mitosis, crossing over does not occur and the resulting replicated chromosome pairs are just that, chromosome pairs. So the tetrad refers solely to a pair of recombined chromosomes during meiosis.

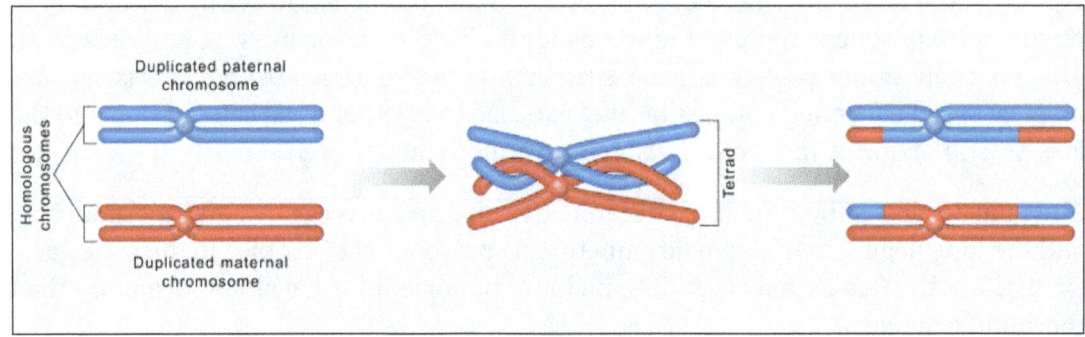

Crossing over – Tetrad formation.

Because tetrads are not exact copies, their development and eventual sharing during reproduction lead to genetic variation across a species. In order to divide, a cell requires a complete set of genetic material in the form of 23 chromatid pairs. Yet meiosis is a two-stage cell division where a single cell forms four daughter cells, each with a single chromosome. In the first stage – meiosis I – a single cell divides into two daughter cells, each containing 23 pairs of chromosomes or 46 chromosomes. In meiosis II, these two daughter cells split again, this time each containing 23 chromosomes, not chromosome pairs. As the only cells that are created by meiosis are sperm and egg cells, this makes sense. During fertilization, a single zygote is formed from two gametes which must contain 23 chromosome pairs – a combination of the genes of both mother and father.

Telophase I

Telophase I is the stage in meiosis I that follows after anaphase I. In anaphase I, the paired homologous chromosomes begin to separate from each other and move towards the opposite ends of the cell. This occurs as the kinetochore microtubules shorten, pulling and separating the paired chromosomes from each other. Telophase I is that phase when the chromosomes have finished moving to opposite ends of the cell. This will then be followed by cytokinesis producing two daughter cells.

After cytokinesis, the two daughter cells would have genetically different chromosomes after meiosis I. Depending on the species, the next stage may be an interkinesis (i.e. a brief phase prior to the second meiotic division) or the daughter cells would immediately proceed to prophase II of meiosis II.

Meiotic telophase I is different from mitotic telophase in a way that the paired chromosomes are the ones that separate from each other in meiotic telophase I and not the sister chromatids as in mitotic telophase. Thus, after meiosis I, each cell would have a haploid set of chromosomes but each chromosome would still be comprised of two sister chromatids.

MEIOSIS 2

Cells move from meiosis I to meiosis II without copying their DNA. Meiosis II is a shorter and simpler process than meiosis I, and you may find it helpful to think of meiosis II as "mitosis for haploid cells."

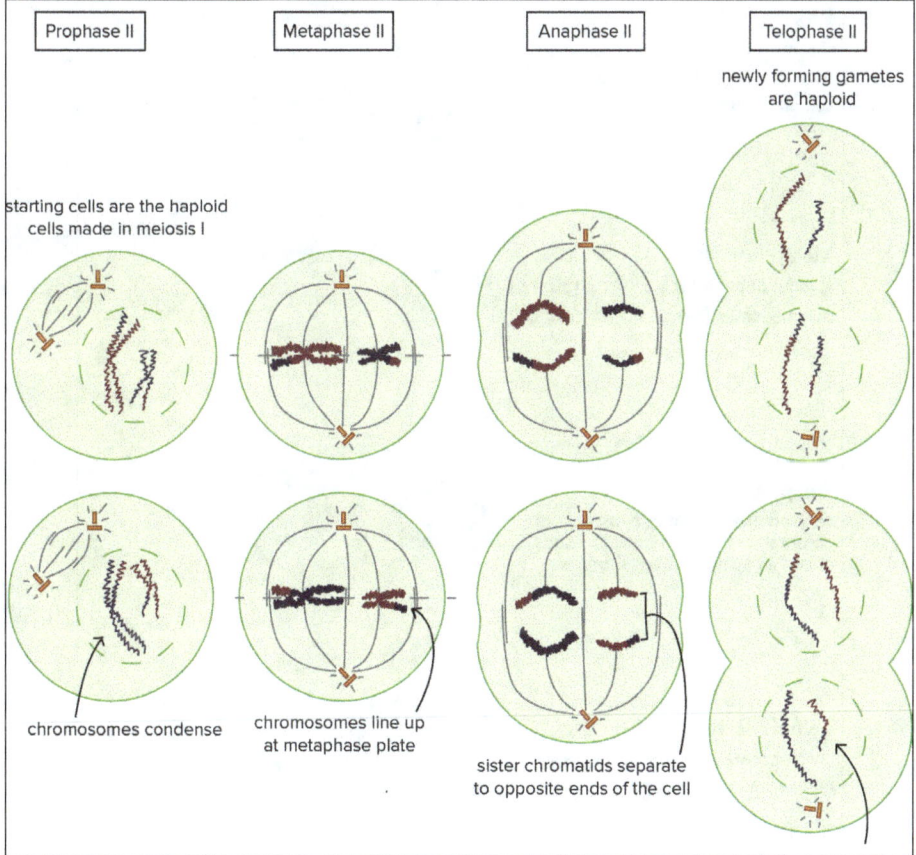

The cells that enter meiosis II are the ones made in meiosis I. These cells are haploid— have just one chromosome from each homologue pair—but their chromosomes still consist of two sister chromatids. In meiosis II, the sister chromatids separate, making haploid cells with non-duplicated chromosomes.

During prophase II, chromosomes condense and the nuclear envelope breaks down, if needed. The centrosomes move apart, the spindle forms between them, and the spindle microtubules begin to capture chromosomes.

The two sister chromatids of each chromosome are captured by microtubules from opposite spindle poles. In metaphase II, the chromosomes line up individually along the metaphase plate. In anaphase II, the sister chromatids separate and are pulled towards opposite poles of the cell.

In telophase II, nuclear membranes form around each set of chromosomes, and the chromosomes decondense. Cytokinesis splits the chromosome sets into new cells, forming the final products of meiosis: four haploid cells in which each chromosome has just one chromatid. In humans, the products of meiosis are sperm or egg cells.

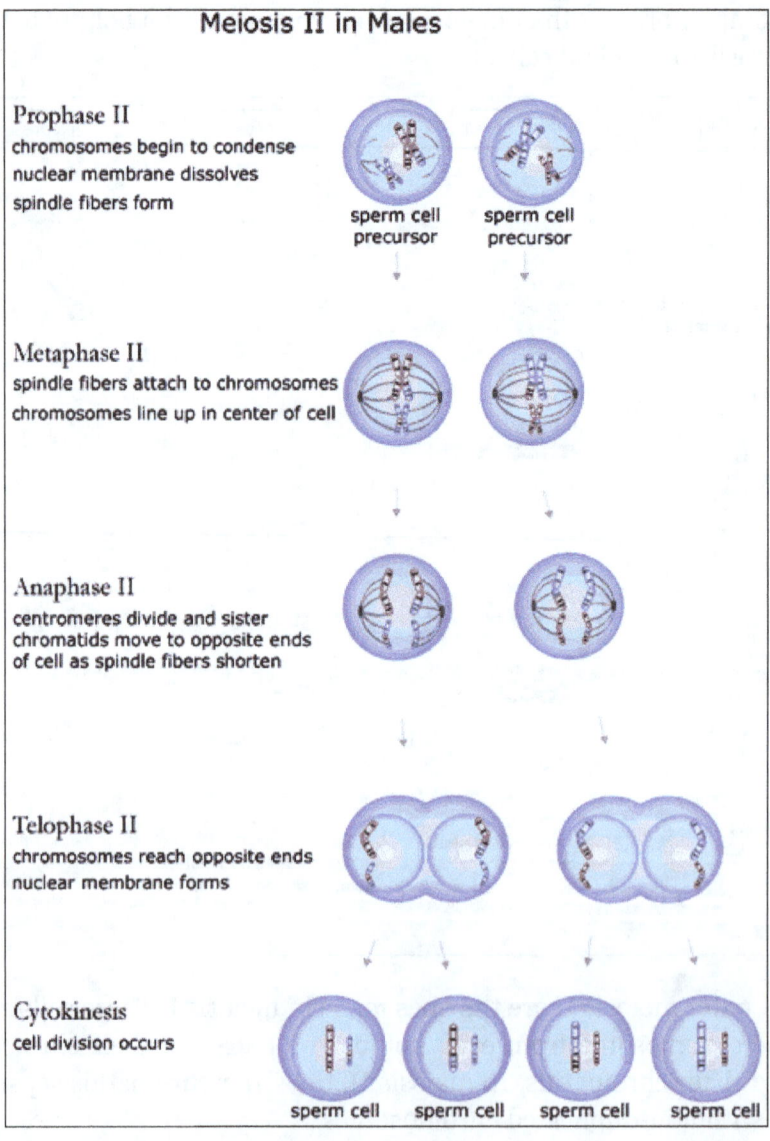

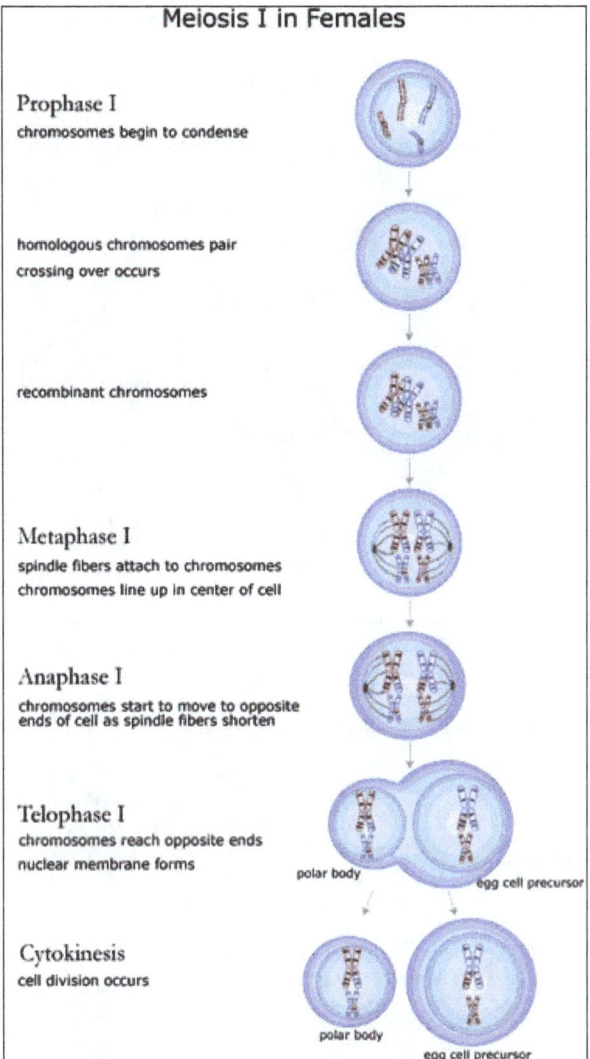

DIFFERENCE BETWEEN MEIOSIS 1 AND MEIOSIS 2

Meiosis I

- It is heterotypic or reduction division.

- The chromosomes remain in the replicated state.

- The number of chromosomes is reduced to half, i.e., from diploid to haploid state.

- Crossing over occurs which makes the two chromatids of a chromosome different.

- It is complicated and long duration division.

- An interphase having both growth phases and synthetic phase precedes meiosis I.

- In prophase I, sister chromatids have convergent arms.

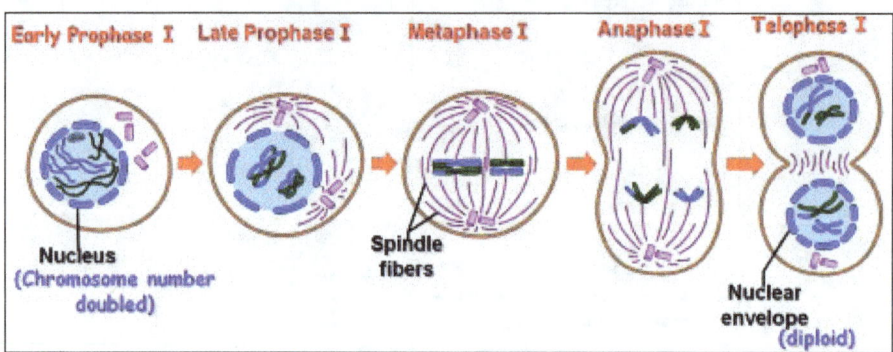

Meiosis II

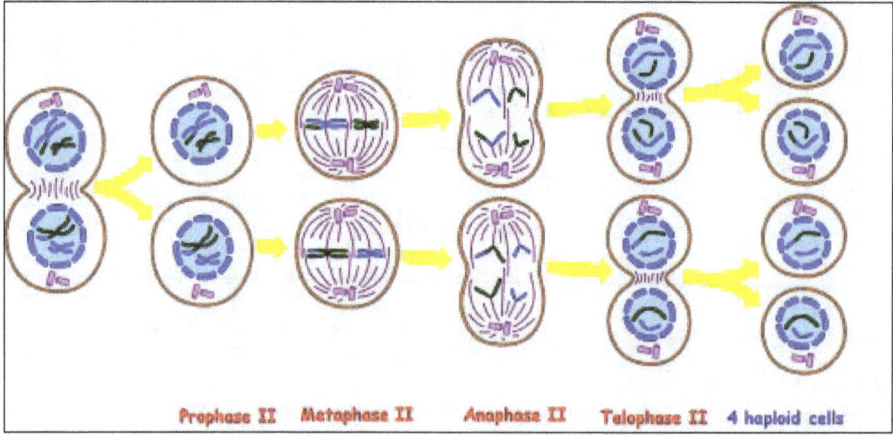

- It is homotypic or equational division.

- The two chromatids of a replicated chromosome separate.

- The number of chromosomes remain the same, i.e., from haploid to haploid state.

- The generally different chromatids of a chromosome are separated.

- It is simple and short duration division.

- The interphase or interkinesis has only growth phase. S phase is absent.

- In prophase II, the sister chromatids have divergent arms.

References

- Cellcycle-mitosis-meiosis, highereducation, vgec: le.ac.uk, Retrieved 29 August, 2020

- Phases-of-meiosis, a, meiosis, cellular-molecular-biology, biology, science: khanacademy.org, Retrieved 2 April, 2020

- Pierce, Benjamin (2009). "Chromossomes and Cell Reproduction". Genetics: A Conceptual Approach, Third Edition. W.H. FREEMAN AND CO. ISBN 9780716779285

- Cellcycle-mitosis-meiosis, highereducation, vgec: le.ac.uk, Retrieved 21 May, 2020

- Difference-between-meiosis-i-and-2013: majordifferences.com, Retrieved 1 February, 2020

5

Genetic Recombination and Crossing Over

Genetic recombination refers to the exchange of genetic material between multiple chromosomes or between different regions of the same chromosome for production of offspring with traits of parents. The exchange of chromosome segments between non-sister chromatids during the production of gametes is termed as crossing over. All the aspects related to genetic recombination and crossing over have been carefully analyzed in this chapter.

RECOMBINATION

Recombination is a primary mechanism through which variation is introduced into populations. Recombination takes place during meiosis, when maternal and paternal genes are regrouped in the formation of gametes (sex cells). Recombination occurs randomly in nature as a normal event of meiosis and is enhanced by the phenomenon of crossing over, in which gene sequences called linkage groups are disrupted, resulting in an exchange of segments between paired chromosomes that are undergoing separation. Thus, although a normal daughter cell produced in meiosis always receives half of the genetic material contained in the parent cell (i.e., is haploid), recombination acts to ensure constant variability: no two daughter cells are identical, nor are any identical in genetic content to the parent cell. Laboratory study of recombination has contributed significantly to the understanding of genetic mechanisms, allowing scientists to map chromosomes, identify linkage groups, isolate the causes of certain genetic anomalies, and manipulates recombination itself by transplantation of genes from one chromosome to another. Because of its potential for creating new—and possibly pathogenic—organisms, experimental recombination can have potentially negative consequences for human health and the health of the environment.

GENETIC RECOMBINATION

Genetic recombination occurs when genetic material is exchanged between two different chromosomes or between different regions within the same chromosome. We can observe it in both eukaryotes (like animals and plants) and prokaryotes (like archaea and bacteria). Keep in mind that in most cases, in order for an exchange to occur, the sequences containing the swapped regions have to be homologous, or similar, to some degree.

The process occurs naturally and can also be carried out in the lab. Recombination increases the genetic diversity in sexually reproducing organisms and can allow an organism to function in new ways.

Examples of Genetic Recombination

Genetic recombination occurs naturally in meiosis. Meiosis is the process of cell division that occurs in eukaryotes, such as humans and other mammals, to produce offspring. In this case, it involves crossing-over. What happens is that two chromosomes, one from each parent, pair up with each other. Next, a segment from one crosses over, or overlaps, a segment of the other. This allows for the swapping of some of their material, as you can see in the illustration below. What we end up with is a new combination of genes that didn't exist before and is not identical to either parent's genetic information. That recombination is also observed in mitosis, but it doesn't occur as often in mitosis as it does in meiosis.

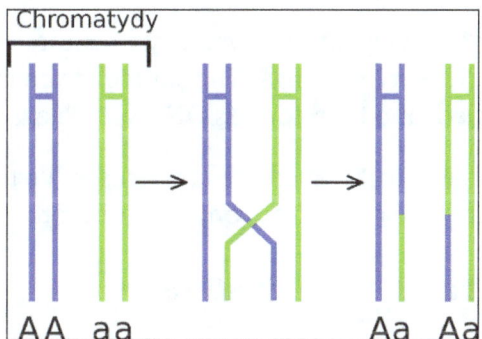

Natural Self-healing

The cell also can also undergo recombinational repair, for example, if it notices that there is a harmful break in the DNA: the kind of break that occurs in both strands. What we observe is an exchange between the broken DNA and a homologous region of DNA that will fill the gaps. There are also other ways that recombination is used to repair DNA.

Types of Genetic Recombination

Scientists have observed the following types of recombination in nature:

- Homologous (general) recombination: As the name implies, this type occurs between DNA molecules of similar sequences. Our cells carry out general recombination during meiosis.

- Nonhomologous (illegitimate) recombination: Again, the name is self-explanatory. This type occurs between DNA molecules that are not necessarily similar. Often, there will be a degree of similarity between the sequences, but it's not as obvious as it would be in homologous recombinations.

- Site-specific recombination: This is observed between particular, very short, sequences, usually containing similarities.

Mitotic recombination: This doesn't actually happen during mitosis, but during interphase, which is the resting phase between mitotic divisions. The process is similar to that in meiotic recombination, and has its possible advantages, but it's usually harmful and can result in tumors. This type of recombination is increased when cells are exposed to radiation.

Prokaryotic cells can undergo recombination through one of these three processes:

- Conjugation is where genes are donated from one organism to another after they have been in contact. At any point, the contact is lost and the genes that were donated to the recipient replace their equivalents in its chromosome. What the offspring ends up having is a mix of traits from different strains of bacteria.

- Transformation: This is where the organism acquires new genes by taking up naked DNA from its surroundings. The source of the free DNA is another bacterium that has died, and therefore its DNA was released to the environment.

Transduction is gene transfer that is mediated by viruses. Viruses called bacteriophages attack bacteria and carry the genes from one bacterium to another.

Homologous (General) Recombination

Homologous recombination is a type of genetic recombination in which nucleotide sequences are exchanged between two similar or identical molecules of double-stranded or single-stranded nucleic acids (usually DNA as in cellular organisms but may be also RNA in viruses). It is most widely used by cells to accurately repair harmful breaks that occur on both strands of DNA, known as double-strand breaks (DSB). Homologous recombination also produces new combinations of DNA sequences during meiosis, the process by which eukaryotes make gamete cells, like sperm and egg cells in animals. These new combinations of DNA represent genetic variation in offspring, which in turn enables populations to adapt during the course of evolution. Homologous recombination is also used in horizontal gene transfer to exchange genetic material between different strains and species of bacteria and viruses.

Although homologous recombination varies widely among different organisms and cell types, for double-stranded DNA (dsDNA) most forms involve the same basic steps. After a double-strand break occurs, sections of DNA around the 5' ends of the break are cut away in a process called resection. In the strand invasion step that follows, an overhanging 3' end of the broken DNA molecule then "invades" a similar or identical DNA molecule that is not broken. After strand invasion, the further sequence of events may follow either of two main pathways discussed below; the DSBR (double-strand break repair) pathway or the SDSA (synthesis-dependent strand annealing) pathway. Homologous recombination that occurs during DNA repair tends to result in non-crossover products, in effect restoring the damaged DNA molecule as it existed before the double-strand break.

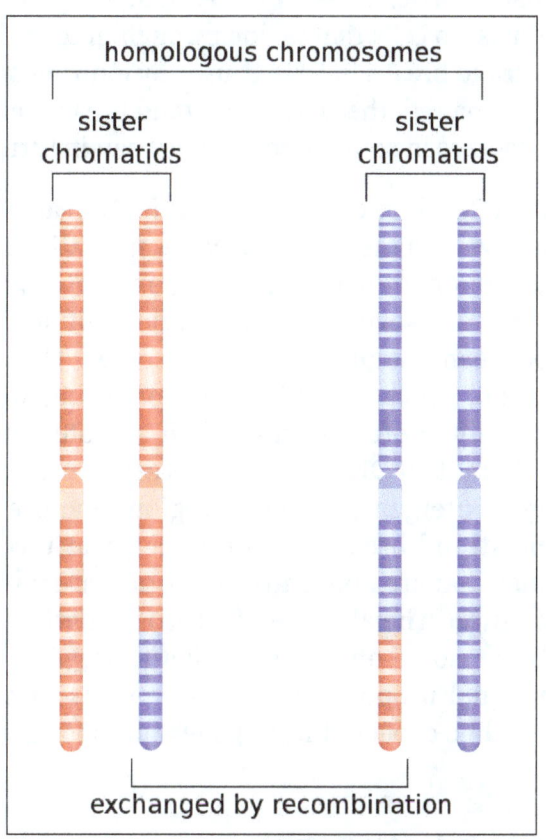

During meiosis, homologous recombination can produce new combinations of genes as shown here between similar but not identical copies of human chromosome 1.

Homologous recombination is conserved across all three domains of life as well as DNA and RNA viruses, suggesting that it is a nearly universal biological mechanism. The discovery of genes for homologous recombination in protists—a diverse group of eukaryotic microorganisms—has been interpreted as evidence that meiosis emerged early in the evolution of eukaryotes. Since their dysfunction has been strongly associated with increased susceptibility to several types of cancer, the proteins that facilitate homologous recombination are topics of active research. Homologous recombination is also used in gene targeting, a technique for introducing genetic changes into target organisms. For their development of this technique, Mario Capecchi, Martin Evans and Oliver Smithies were awarded the 2007 Nobel Prize for Physiology or Medicine; Capecchi and Smithies independently discovered applications to mouse embryonic stem cells, however the highly conserved mechanisms underlying the DSB repair model, including uniform homologous integration of transformed DNA (gene therapy), were first shown in plasmid experiments by Orr-Weaver, Szostack and Rothstein. Researching the plasmid-induced DSB, using γ-irradiation in the 1970s-1980s, led to later experiments using endonucleases (e.g. I-SceI) to cut chromosomes for genetic engineering of mammalian cells, where nonhomologous recombination is more frequent than in yeast.

Homologous recombination (HR) is essential to cell division in eukaryotes like plants, animals, fungi and protists. In cells that divide through mitosis, homologous recombination repairs double-strand breaks in DNA caused by ionizing radiation or DNA-damaging chemicals. Left unrepaired, these double-strand breaks can cause large-scale rearrangement of chromosomes in somatic cells, which can in turn lead to cancer.

In addition to repairing DNA, homologous recombination also helps produce genetic diversity when cells divide in meiosis to become specialized gamete cells—sperm or egg cells in animals, pollen or ovules in plants, and spores in fungi. It does so by facilitating chromosomal crossover, in which regions of similar but not identical DNA are exchanged between homologous chromosomes. This creates new, possibly beneficial combinations of genes, which can give offspring an evolutionary advantage. Chromosomal crossover often begins when a protein called Spo11 makes a targeted double-strand break in DNA. These sites are non-randomly located on the chromosomes; usually in intergenic promoter regions and preferentially in GC-rich domains These double-strand break sites often occur at recombination hotspots, regions in chromosomes that are about 1,000–2,000 base pairs in length and have high rates of recombination. The absence of a recombination hotspot between two genes on the same chromosome often means that those genes will be inherited by future generations in equal proportion. This represents linkage between the two genes greater than would be expected from genes that independently assort during meiosis.

Timing within the Mitotic Cell Cycle

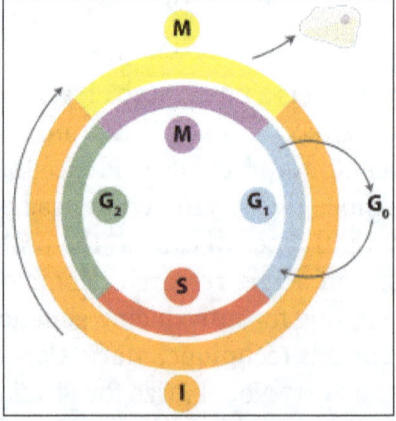

Homologous recombination repairs DNA before the cell enters mitosis (M phase). It occurs only during and shortly after DNA replication, during the S and G_2 phases of the cell cycle.

Double-strand breaks can be repaired through homologous recombination or through non-homologous end joining (NHEJ). NHEJ is a DNA repair mechanism which, unlike homologous recombination, does not require a long homologous sequence to guide repair. Whether homologous recombination or NHEJ is used to repair double-strand breaks is largely determined by the phase of cell cycle. Homologous recombination repairs DNA before the cell enters mitosis (M phase). It occurs during and shortly after DNA replication, in the S and G_2 phases of the cell cycle, when sister chromatids are more easily available. Compared to homologous chromosomes, which are similar to another chromosome but often have different alleles, sister chromatids are an ideal template for homologous recombination because they are an identical copy of a given chromosome. In contrast to homologous recombination, NHEJ is predominant in the G_1 phase of the cell cycle, when the cell is growing but not yet ready to divide. It occurs less frequently after the G_1 phase, but maintains at least some activity throughout the cell cycle. The mechanisms that regulate homologous recombination and NHEJ throughout the cell cycle vary widely between species.

Cyclin-dependent kinases (CDKs), which modify the activity of other proteins by adding phosphate groups to (that is, phosphorylating) them, are important regulators of homologous recombination in eukaryotes. When DNA replication begins in budding yeast, the cyclin-dependent kinase Cdc28 begins homologous recombination by phosphorylating the Sae2 protein. After being so activated by the addition of a phosphate, Sae2 uses its endonuclease activity to make a clean cut near a double-strand break in DNA. This allows a three-part protein known as the MRX complex to bind to DNA, and begins a series of protein-driven reactions that exchange material between two DNA molecules.

Preliminary Steps

The packaging of eukaryotic DNA into chromatin presents a barrier to all DNA-based processes that require recruitment of enzymes to their sites of action. To allow HR DNA repair, the chromatin must be remodeled. In eukaryotes, ATP dependent chromatin remodeling complexes and histone-modifying enzymes are two predominant factors employed to accomplish this remodeling process.

Chromatin relaxation occurs rapidly at the site of DNA damage. In one of the earliest steps, the stress-activated protein kinase, c-Jun N-terminal kinase (JNK), phosphorylates SIRT6 on serine 10 in response to double-strand breaks or other DNA damage. This post-translational modification facilitates the mobilization of SIRT6 to DNA damage sites, and is required for efficient recruitment of poly (ADP-ribose) polymerase 1 (PARP1) to DNA break sites and for efficient repair of DSBs. PARP1 protein starts to appear at DNA damage sites in less than a second, with half maximum accumulation within 1.6 seconds after the damage occurs. Next the chromatin remodeler Alc1 quickly attaches to the product of PARP1 action, a poly-ADP ribose chain, and Alc1 completes arrival at the DNA damage within 10 seconds of the occurrence of the damage. About half of the maximum chromatin relaxation, presumably due to action of Alc1, occurs by 10 seconds. This then allows recruitment of the DNA repair enzyme MRE11, to initiate DNA repair, within 13 seconds.

γH2AX, the phosphorylated form of H2AX is also involved in the early steps leading to chromatin decondensation after DNA double-strand breaks. The histone variant H2AX constitutes about 10% of the H2A histones in human chromatin. γH2AX (H2AX phosphorylated on serine 139) can be detected as soon as 20 seconds after irradiation of cells (with DNA double-strand break formation), and half maximum accumulation of γH2AX occurs in one minute. The extent of chromatin with phosphorylated γH2AX is about two million base pairs at the site of a DNA double-strand break. γH2AX does not, itself, cause chromatin decondensation, but within 30 seconds of irradiation, RNF8 protein can be detected in association with γH2AX. RNF8 mediates extensive chromatin decondensation, through its subsequent interaction with CHD4, a component of the nucleosome remodeling and deacetylase complex NuRD. After undergoing relaxation subsequent to DNA damage, followed by DNA repair, chromatin recovers to a compaction state close to its pre-damage level after about 20 min.

Models

Two primary models for how homologous recombination repairs double-strand breaks in DNA are the double-strand break repair (DSBR) pathway (sometimes called the double Holliday junction model) and the synthesis-dependent strand annealing (SDSA) pathway. The two pathways are similar in their first several steps. After a double-strand break occurs, the MRX complex (MRN complex in humans) binds to DNA on either side of the break. Next a resection, in which DNA around the 5' ends of the break is cut back, is carried out in two distinct steps. In the first step of resection, the MRX complex recruits the Sae2 protein. The two proteins then trim back the 5' ends on either side of the break to create short 3' overhangs of single-strand DNA. In the second step, 5'→3' resection is continued by the Sgs1 helicase and the Exo1 and Dna2 nucleases. As a helicase, Sgs1 "unzips" the double-strand DNA, while Exo1 and Dna2's nuclease activity allows them to cut the single-stranded DNA produced by Sgs1.

The RPA protein, which has high affinity for single-stranded DNA, then binds the 3' overhangs. With the help of several other proteins that mediate the process, the Rad51 protein (and Dmc1, in meiosis) then forms a filament of nucleic acid and protein on the single strand of DNA coated with RPA. This nucleoprotein filament then begins searching for DNA sequences similar to that of the 3' overhang. After finding such a sequence, the single-stranded nucleoprotein filament moves into (invades) the similar or identical recipient DNA duplex in a process called strand invasion. In cells that divide through mitosis, the recipient DNA duplex is generally a sister chromatid, which is identical to the damaged DNA molecule and provides a template for repair. In meiosis, however, the recipient DNA tends to be from a similar but not necessarily identical homologous chromosome. A displacement loop (D-loop) is formed during strand invasion between the invading 3' overhang strand and the homologous chromosome. After strand invasion, a DNA polymerase extends the end of the invading 3' strand by synthesizing new DNA. This changes the D-loop to a cross-shaped structure known as a Holliday junction. Following this, more DNA synthesis occurs on the invading strand (i.e., one of the original 3' overhangs), effectively restoring the strand on the homologous chromosome that was displaced during strand invasion.

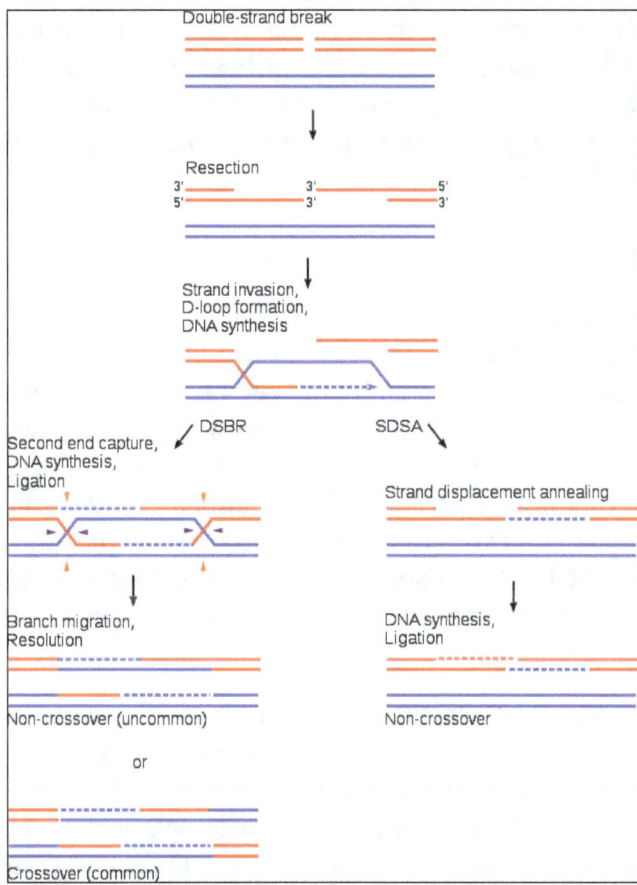

The DSBR and SDSA pathways follow the same initial steps, but diverge thereafter. The DSBR pathway most often results in chromosomal crossover (bottom left), while SDSA always ends with non-crossover products (bottom right).

DSBR Pathway

After the stages of resection, strand invasion and DNA synthesis, the DSBR and SDSA pathways become distinct. The DSBR pathway is unique in that the second 3' overhang (which was not involved in strand invasion) also forms a Holliday junction with the homologous chromosome. The double Holliday junctions are then converted into recombination products by nicking endonucleases, a type of restriction endonuclease which cuts only one DNA strand. The DSBR pathway commonly results in crossover, though it can sometimes result in non-crossover products; the ability of a broken DNA molecule to collect sequences from separated donor loci was shown in mitotic budding yeast using plasmids or endonuclease induction of chromosomal events. Because of this tendency for chromosomal crossover, the DSBR pathway is a likely model of how crossover homologous recombination occurs during meiosis.

Whether recombination in the DSBR pathway results in chromosomal crossover is determined by how the double Holliday junction is cut, or "resolved". Chromosomal crossover will occur if one Holliday junction is cut on the crossing strand and the other Holliday junction is cut on the non-crossing strand (in figure, along the horizontal purple arrowheads at one Holliday junction and along the vertical orange arrowheads at the other). Alternatively, if the two Holliday junctions are cut on the crossing strands (along the horizontal purple arrowheads at both Holliday junctions in figure), then chromosomes without crossover will be produced.

SDSA Pathway

Homologous recombination via the SDSA pathway occurs in cells that divide through mitosis and meiosis and results in non-crossover products. In this model, the invading 3' strand is extended along the recipient DNA duplex by a DNA polymerase, and is released as the Holliday junction between the donor and recipient DNA molecules slides in a process called branch migration. The newly synthesized 3' end of the invading strand is then able to anneal to the other 3' overhang in the damaged chromosome through complementary base pairing. After the strands anneal, a small flap of DNA can sometimes remain. Any such flaps are removed, and the SDSA pathway finishes with the resealing, also known as ligation, of any remaining single-stranded gaps.

During mitosis, the major homologous recombination pathway for repairing DNA double-strand breaks appears to be the SDSA pathway (rather than the DSBR pathway). The SDSA pathway produces non-crossover recombinants. During meiosis non-crossover recombinants also occur frequently and these appear to arise mainly by the SDSA pathway as well. Non-crossover recombination events occurring during meiosis likely reflect instances of repair of DNA double-strand damages or other types of DNA damages.

SSA Pathway

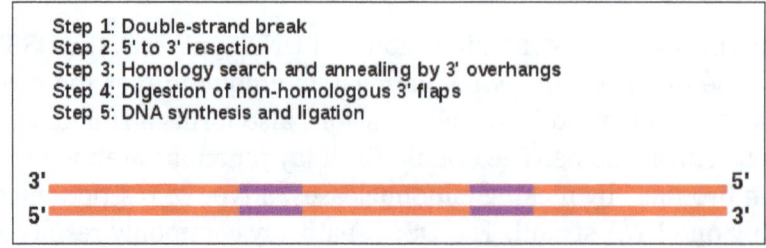

Recombination via the SSA pathway occurs between two repeat elements (purple) on the same DNA duplex, and results in deletions of genetic material.

The single-strand annealing (SSA) pathway of homologous recombination repairs double-strand breaks between two repeat sequences. The SSA pathway is unique in that it does not require a separate similar or identical molecule of DNA, like the DSBR or SDSA pathways of homologous recombination. Instead, the SSA pathway only requires a single DNA duplex, and uses the repeat sequences as the identical sequences that homologous recombination needs for repair. The pathway is relatively simple in concept: After two strands of the same DNA duplex are cut back around the site of the double-strand break, the two resulting 3' overhangs then align and anneal to each other, restoring the DNA as a continuous duplex.

As DNA around the double-strand break is cut back, the single-stranded 3' overhangs being produced are coated with the RPA protein, which prevents the 3' overhangs from sticking to themselves. A protein called Rad52 then binds each of the repeat sequences on either side of the break, and aligns them to enable the two complementary repeat sequences to anneal. After annealing is complete, leftover non-homologous flaps of the 3' overhangs are cut away by a set of nucleases, known as Rad1/Rad10, which are brought to the flaps by the Saw1 and Slx4 proteins. New DNA synthesis fills in any gaps, and ligation restores the DNA duplex as two continuous strands. The DNA sequence between the repeats is always lost, as is one of the two repeats. The SSA pathway is considered mutagenic since it results in such deletions of genetic material.

BIR Pathway

During DNA replication, double-strand breaks can sometimes be encountered at replication forks as DNA helicase unzips the template strand. These defects are repaired in the break-induced replication (BIR) pathway of homologous recombination. The precise molecular mechanisms of the BIR pathway remain unclear. Three proposed mechanisms have strand invasion as an initial step, but they differ in how they model the migration of the D-loop and later phases of recombination.

The BIR pathway can also help to maintain the length of telomeres (regions of DNA at the end of eukaryotic chromosomes) in the absence of (or in cooperation with) telomerase. Without working copies of the enzyme telomerase, telomeres typically shorten with each cycle of mitosis, which eventually blocks cell division and leads to senescence. In budding yeast cells where telomerase has been inactivated through mutations, two types of "survivor" cells have been observed to avoid senescence longer than expected by elongating their telomeres through BIR pathways.

Maintaining telomere length is critical for cell immortalization, a key feature of cancer. Most cancers maintain telomeres by upregulating telomerase. However, in several types of human cancer, a BIR-like pathway helps to sustain some tumors by acting as an alternative mechanism of telomere maintenance. This fact has led scientists to investigate whether such recombination-based mechanisms of telomere maintenance could thwart anti-cancer drugs like telomerase inhibitors.

Effects of Dysfunction

Without proper homologous recombination, chromosomes often incorrectly align for the first phase of cell division in meiosis. This causes chromosomes to fail to properly segregate in a process called nondisjunction. In turn, nondisjunction can cause sperm and ova to have too few or too many chromosomes. Down's syndrome, which is caused by an extra copy of chromosome 21, is one of many abnormalities that result from such a failure of homologous recombination in meiosis.

Deficiencies in homologous recombination have been strongly linked to cancer formation in humans. For example, each of the cancer-related diseases Bloom's syndrome, Werner's syndrome and Rothmund-Thomson syndrome are caused by malfunctioning copies of RecQ helicase genes involved in the regulation of homologous recombination: BLM, WRN and RECQL4, respectively. In the cells of Bloom's syndrome patients, who lack a working copy of the BLM protein, there is an elevated rate of homologous recombination. Experiments in mice deficient in BLM have suggested that the mutation gives rise to cancer through a loss of heterozygosity caused by increased homologous recombination. A loss in heterozygosity refers to the loss of one of two versions—or alleles—of a gene. If one of the lost alleles helps to suppress tumors, like the gene for the retinoblastoma protein for example, then the loss of heterozygosity can lead to cancer.

Decreased rates of homologous recombination cause inefficient DNA repair, which can also lead to cancer. This is the case with BRCA1 and BRCA2, two similar tumor suppressor genes whose malfunctioning has been linked with considerably increased risk for breast and ovarian cancer. Cells missing BRCA1 and BRCA2 have a decreased rate of homologous recombination and increased sensitivity to ionizing radiation, suggesting that decreased homologous recombination leads to increased susceptibility to cancer. Because the only known function of BRCA2 is to help initiate homologous recombination, researchers have speculated that more detailed knowledge of BRCA2's role in homologous recombination may be the key to understanding the causes of breast and ovarian cancer. Tumours with a homologous recombination deficiency (including BRCA defects) are described as HRD-positive.

Evolutionary Conservation

While the pathways can mechanistically vary, the ability of organisms to perform homologous recombination is universally conserved across all domains of life. Based on the similarity of their amino acid sequences, homologs of a number of proteins can be found in multiple domains of life indicating that they evolved a long time ago, and have since diverged from common ancestral proteins.

RecA recombinase family members are found in almost all organisms with RecA in bacteria, Rad51 and DMC1 in eukaryotes, RadA in archaea, and UvsX in T4 phage. Related single stranded binding proteins that are important for homologous recombination, and many other processes, are also found in all domains of life. Rad54, Mre11, Rad50, and a number of other proteins are also found in both archaea and eukaryotes.

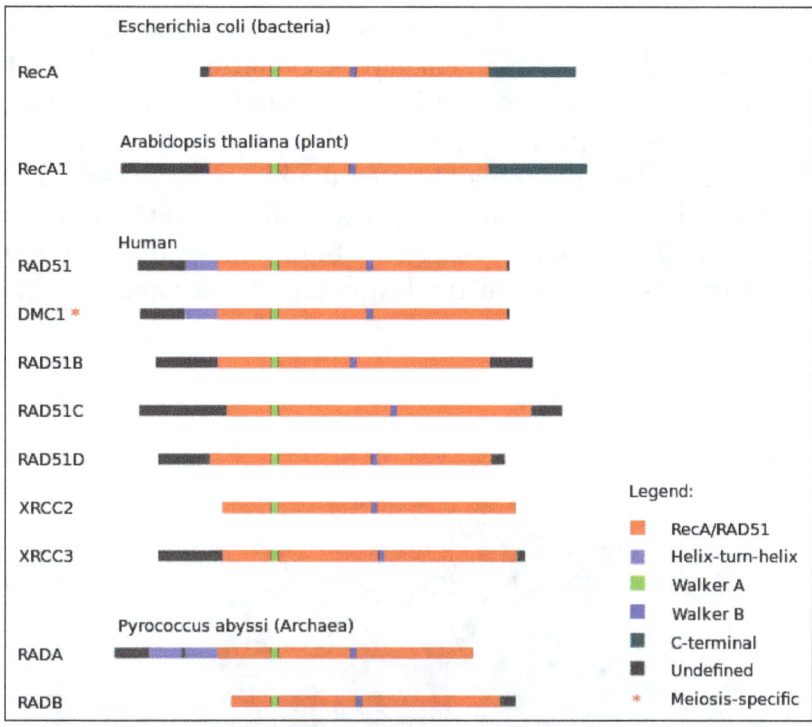

Protein domains in homologous recombination-related proteins are conserved across the three main groups of life: Archaea, bacteria and eukaryotes.

RecA Recombinase Family

The proteins of the RecA recombinase family of proteins are thought to be descended from a common ancestral recombinase. The RecA recombinase family contains RecA protein from bacteria, the Rad51 and Dmc1 proteins from eukaryotes, and RadA from archaea, and the recombinase paralog proteins. Studies modeling the evolutionary relationships between the Rad51, Dmc1 and RadA proteins indicate that they are monophyletic, or that they share a common molecular ancestor. Within this protein family, Rad51 and Dmc1 are grouped together in a separate clade from RadA. One of the reasons for grouping these three proteins together is that they all possess a modified helix-turn-helix motif, which helps the proteins bind to DNA, toward their N-terminal ends. An ancient gene duplication event of a eukaryotic RecA gene and subsequent mutation has been proposed as a likely origin of the modern RAD51 and DMC1 genes.

The proteins generally share a long conserved region known as the RecA/Rad51 domain. Within this protein domain are two sequence motifs, Walker A motif and Walker B motif. The Walker A and B motifs allow members of the RecA/Rad51 protein family to engage in ATP binding and ATP hydrolysis.

Meiosis-specific Proteins

The discovery of Dmc1 in several species of Giardia, one of the earliest protists to diverge as a eukaryote, suggests that meiotic homologous recombination—and thus meiosis itself—emerged very early in eukaryotic evolution. In addition to research on Dmc1, studies on the Spo11 protein have provided information on the origins of meiotic recombination. Spo11, a type II topoisomerase, can initiate homologous recombination in meiosis by making targeted double-strand breaks in DNA. Phylogenetic trees based on the sequence of genes similar to SPO11 in animals, fungi, plants, protists and archaea have led scientists to believe that the version Spo11 currently in eukaryotes emerged in the last common ancestor of eukaryotes and archaea.

Technological Applications

Gene Targeting

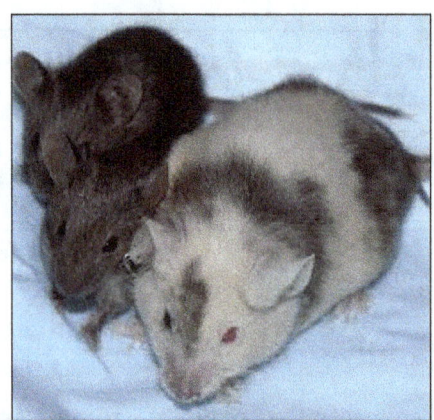

As a developing embryo, this chimeric mouse had the agouti coat color gene introduced into its DNA via gene targeting. Its offspring are homozygous for the agouti gene.

Many methods for introducing DNA sequences into organisms to create recombinant DNA and genetically modified organisms use the process of homologous recombination. Also called gene targeting, the method is especially common in yeast and mouse genetics. The gene targeting method in knockout mice uses mouse embryonic stem cells to deliver artificial genetic material (mostly of therapeutic interest), which represses the target gene of the mouse by the principle of homologous recombination. The mouse thereby acts as a working model to understand the effects of a specific mammalian gene. In recognition of their discovery of how homologous recombination can be used to introduce genetic modifications in mice through embryonic stem cells, Mario Capecchi, Martin Evans and Oliver Smithies were awarded the 2007 Nobel Prize for Physiology or Medicine.

Advances in gene targeting technologies which hijack the homologous recombination mechanics of cells are now leading to the development of a new wave of more accurate, isogenic human disease models. These engineered human cell models are thought to more accurately reflect the genetics of human diseases than their mouse model predecessors. This is largely because mutations of interest are introduced into endogenous genes, just as they occur in the real patients, and because they are based on human genomes rather than rat genomes. Furthermore, certain technologies enable the knock-in of a particular mutation rather than just knock-outs associated with older gene targeting technologies.

Protein Engineering

Protein engineering with homologous recombination develops chimeric proteins by swapping fragments between two parental proteins. These techniques exploit the fact that recombination can introduce a high degree of sequence diversity while preserving a protein's ability to fold into its tertiary structure, or three-dimensional shape. This stands in contrast to other protein engineering techniques, like random point mutagenesis, in which the probability of maintaining protein function declines exponentially with increasing amino acid substitutions. The chimeras produced by recombination techniques are able to maintain their ability to fold because their swapped parental fragments are structurally and evolutionarily conserved. These recombinable "building blocks" preserve structurally important interactions like points of physical contact between different amino acids in the protein's structure. Computational methods like SCHEMA and statistical coupling analysis can be used to identify structural subunits suitable for recombination.

Techniques that rely on homologous recombination have been used to engineer new proteins. researchers were able to create chimeras of two enzymes involved in the biosynthesis of isoprenoids, a diverse class of compounds including hormones, visual pigments and certain pheromones. The chimeric proteins acquired an ability to catalyze an essential reaction in isoprenoid biosynthesis—one of the most diverse pathways of biosynthesis found in nature—that was absent in the parent proteins. Protein engineering through recombination has also produced chimeric enzymes with new function in members of a group of proteins known as the cytochrome P450 family, which in humans is involved in detoxifying foreign compounds like drugs, food additives and preservatives.

Cancer Therapy

Cancer cells with BRCA mutations have deficiencies in homologous recombination, and drugs to exploit those deficiencies have been developed and used successfully in clinical trials. Olaparib, a PARP1 inhibitor, shrunk or stopped the growth of tumors from breast, ovarian and prostate cancers caused by mutations in the BRCA1 or BRCA2 genes, which are necessary for HR. When BRCA1 or BRCA2 is absent, other types of DNA repair mechanisms must compensate for the deficiency of HR, such as base-excision repair (BER) for stalled replication forks or non-homologous end joining (NHEJ) for double strand breaks. By inhibiting BER in an HR-deficient cell, olaparib applies the concept of synthetic lethality to specifically target cancer cells. While PARP1 inhibitors represent a novel approach to cancer therapy, researchers have cautioned that they may prove insufficient for treating late-stage metastatic cancers. Cancer cells can become resistant to a PARP1 inhibitor if they undergo deletions of mutations in BRCA2, undermining the drug's synthetic lethality by restoring cancer cells' ability to repair DNA by HR.

Nonhomologous (Illegitimate) Recombination

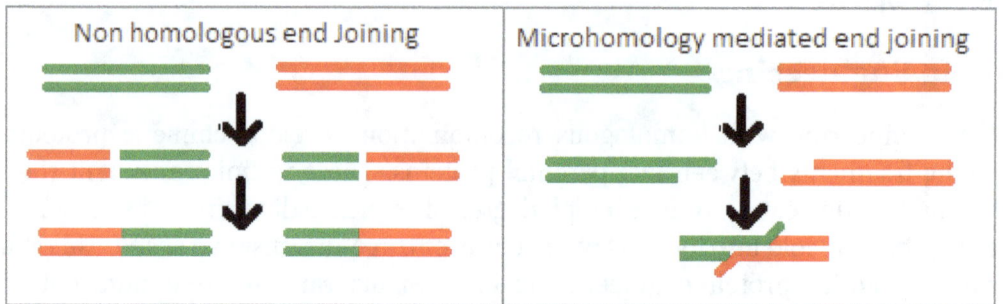

Non-homologous end joining a process of illegitimate recombination versus a homology driven recombination event.

Illegitimate recombination, or nonhomologous recombination, is the process by which two unrelated double stranded segments of DNA are joined. This insertion of genetic material which is not meant to be adjacent tends to lead to genes being broken causing the protein which they encode to not be properly expressed. One of the primary pathways by which this will occur is the repair mechanism known as Non-homologous end joining (NHEJ).

Illegitimate recombination is a natural process which was first found to be present within E. coli. A 700-1400 base pair segment of DNA was found to have inserted itself into the gal and lac operons resulting in a strong polar mutation. This mechanism was then found to have the ability to insert other short genetic sequences into other locations within the bacterial genome often leading to a change in the expression of neighboring genes. Oftentimes it leads to the neighboring genes to simply shut off. However some of these segments also had strong start and stop signals which changed the regulation of neighboring genes leading in changes in the amount of transcription. What differentiated this form of genetic recombination from those dependent of genetic homology was that the process observed as illegitimate did not require the use of homologous segments of DNA. While not being entirely understood at the time, it was recognized to hold potential in generating changes in the chromosomal evolution.

Mechanism

The mechanism of Illegitimate recombination is that of non-homologous end joining in which two strands of DNA not sharing homology will be joined together by the gene repair machinery. Upon recognition of a double strand break a protein complex will keep the two strands within close enough proximity in order to allow for repair of the strands. Next the ends of the DNA are repaired such that any incorrect or damaged nucleotides are removed. Once this happens the strands are able to be ligated together such that they are a single strand of DNA which previously had not been adjacent. This process is common for eukaryotic cells and tends to act as a repair mechanism, but can lead to these mutations if illegitimate recombination occurs. The illegitimate recombination will often take the form of large chromosomal aberrations within a eukaryotic organism as it has much larger segments of DNA than prokaryotic cells. As of such non-homologous end joining can cause illegitimate recombination which creates insertion and deletion mutations in chromosomes as well as translocation of one chromosomal segment to that of another chromosome. These large scale changes in the chromosome in eukaryotic organisms tend to have deleterious effects on the organism rather than conferring a type of genetic advantage.

Deleterious Effects on Organisms

Illegitimate recombination often times has deleterious effects on an organism as it results in a large scale change on the genetic sequence of an organism. These changes will result in mutations as the joining of DNA not based on homology will most often place genetic elements in locations in which they previously had not been placed. This can disrupt the function of genes which may be essential to the function of an organism. In the case of cancer it has been found that tumors can be a result of illegitimate recombination resulting in hairpin formation which alter the gene function within the genome of tumor cells.

Applications

Illegitimate recombination is a tool which can be used in the laboratory as well as it is a useful research tool. Illegitimate recombination can generate random mutagenesis in order to generate a random alteration of the genetic sequence of an organism. The induction of this mutagenesis allows for the study of a genetic sequence by creating a mutation in a genetic segment altering the function of that genetic segment. This allows for the study of gene function through the analysis of differences between mutants and natural organisms to interpret what process a gene is linked to.

Advantages of Genetic Recombination

Recombination has at least two additional benefits for sexual species. It makes new combinations of alleles along chromosomes, and it restricts the effects of mutations largely to the region around a gene, not the whole chromosome.

Since each chromosome undergoes at least one recombination event during meiosis, new combinations of alleles are generated. The arrangement of alleles inherited from each parent are not preserved, but rather the new germ cells carry chromosomes with new combinations of alleles of the genes. This remixing of combinations of alleles is a rich source of diversity in a population.

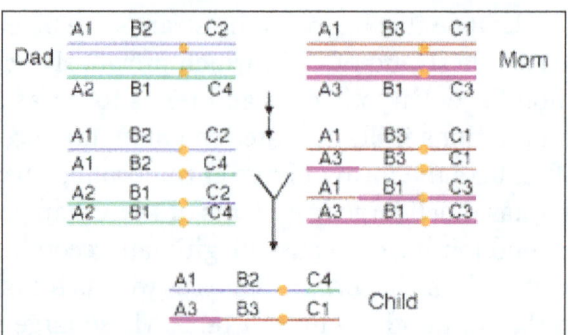

Recombination during meiosis generates new combinations of alleles in the off-spring. One homologous pair of chromosomes is illustrated, starting at the "four-strand" stage. Each line is a duplex DNA molecule in a chromatid. The two chromosomes in the father (inherited from the paternal grandparents) are blue and green; the homologous chromosomes in the mother (inherited from the maternal grandparents) are brown and pink. All chromosomes have genes A, B and C; different numbers refer to different alleles. In this illustration, a crossover on the short arm of the chromosome during development of the male germ cells links allele 4 of gene C with alleles 1 of gene A and allele 2 of gene B, as well as the reciprocal arrangement. A crossover on the long arm of the chromosome is illustrated for development of the female germ cell, making the new combination A3, B3 and C1. A child can have the new chromosomes A1B2C4 and A3B3C1.Tthat neither of these combinations was in the father or mother.

Over time, recombination will separate alleles at one locus from alleles at a linked locus. A chromosome through generations is not fixed, but rather it is "fluid," having many different combinations of alleles. This allows non-functional (less functional) alleles to be cleared from a population. If recombination did not occur, then one deleterious mutant allele would cause an entire chromosome to be eliminated from the population. However, with recombination, the mutant allele can be separated from the other genes on that chromosome. Then negative selection can remove defective alleles of a gene from a population while affecting the frequency of alleles only of genes in tight linkage to the mutant gene. Conversely, the rare beneficial alleles of genes can be tested in a population without being irreversibly linked to any potentially deleterious mutant alleles of nearby genes. This keeps the effective target size for mutation close to that of a gene, not the whole chromosome.

RECOMBINATION IN MITOSIS AND MEIOSIS

Mitotic Recombination

A diploid organism has two copies of each chromosome. If it has four chromosomes, there are two pairs, A and A' and B and B', not four different chromosomes A, B, C and

D. One copy of each chromosome came from its father (e.g. A and B) and one copy of each came from its mother (e.g. A' and B'). Meiosis is the process of reductive division whereby a diploid organism generates haploid germ cells (in this case, with two chromosomes), and each germ cell has a single copy of each chromosome. In this example, meiosis does not generate germ cells with A and A' or B and B', rather it produces cells with A and B, or A and B', or A' and B, or A' and B'. The homologous chromosomes, each consisting of two sister chromatids, are paired during the first phase of meiosis, e.g., A with A' and B with B'. Then the homologous chromosomes are moved to separate cells at the end of the first phase, insuring that the two homologs do not stay together during reductive division in the second phase of meiosis. Thus each germ cell receives the haploid complement of the genetic material, i.e. one copy of each chromosome. The combination of two haploid sets of chromosomes during fertilization restores the diploid state, and the cycle can resume. Failure to distribute one copy of each chromosome to each germ cell has severe consequences. Absence of one copy of a chromosome in an otherwise diploid zygote is likely fatal. Having an extra copy of a chromosome (trisomy) also causes problems. In humans, trisomy for chromosomes 15 or 18 results in perinatal death and trisomy 21 leads to developmental defects known as Down's syndrome.

The ability of homologous chromosomes to be paired during the first phase of meiosis is fundamental to the success of this process, which maintains a correct haploid set of chromosomes in the germ cell. Recombination is an integral part of the pairing of homologous chromosomes. It occurs between non-sister chromatids during the pachytene stage of meiosis I (the first stage of meiosis) and possibly before, when the homologous chromosomes are aligned in zygotene. The crossovers of recombination are visible in the diplotene phase. During this phase, the homologous chromosomes partially separate, but they are still held together at joints called chiasmata; these are likely the actual crossovers between chromatids of homologous chromosomes. The chiasmata are progressively broken as meiosis I is completed, corresponding to resolution of the recombination intermediates. During anaphase and telophase of meiosis I, each homologous chromosome moves to a different cell, i.e. A and A' in different cells, B and B' in different cells in the example. Thus recombinations occur in every meiosis, resulting in at least one exchange between pairs of homologous chromosomes per meiosis.

Recent genetic evidence demonstrates that recombination is required for homologous pairing of chromosomes during meiosis. Genetic screens have revealed mutants of yeast and Drosophila that block pairing of homologous chromosomes. These are also defective in recombination. Likewise, mutants defective in some aspects of recombination are also defective in pairing. Indeed, the process of synapsis (or pairing) between homologous chromosomes in zygotene, crossing over between homologs in pachytene, and resolution of the crossovers in the latter phases of meiosis I (diakinesis, metaphase I, and anaphase I) correspond to the synapsis, formation of a recombinant joint and resolution that mark the progression of recombination.

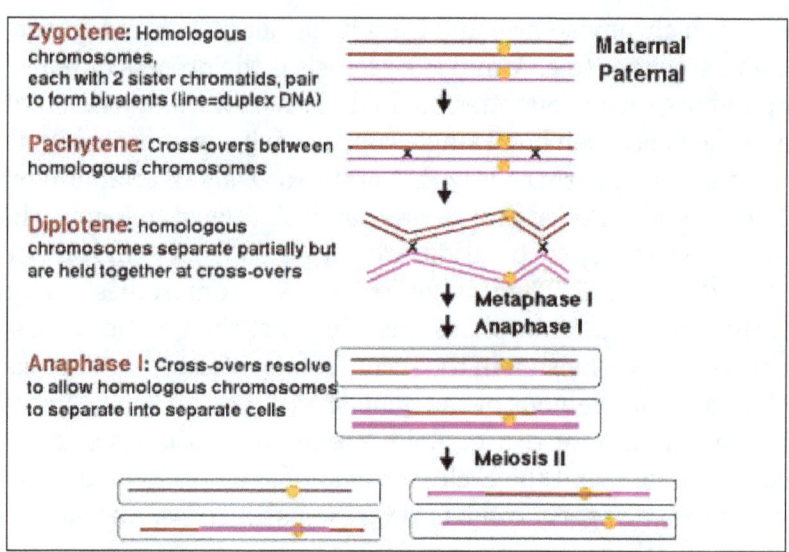

Homologous pairing and recombination during the first stage of meiosis (meiosis I). After DNA synthesis has been completed, two copies of each homologous chromosome are still connected at centromeres (yellow circles). This diagram starts with replicated chromosomes, referred to as the four-strand stage in the literature on meiosis and recombination. In this usage, each "strand" is a chromatid and is a duplex DNA molecule. In this diagram, each duplex DNA molecule is shown as a single line, brown for the two sister chromatids of chromosome derived from the mother (maternal) and pink for the sister chromatids from the paternal chromosome. Only one homologous pair is shown, but usually there are many more, e.g. 4 pairs of chromosomes in Drosophilaand 23 pairs in humans. During the meiosis I, the homologous chromosomes align and then separate. At the zygotene stage, the two homologous chromosomes, each with two sister chromatids, pair along their length in a process called synapsis. The resulting group of four chromatids is called a tetrad or bivalent. During pachytene, recombination occurs between a maternal and a paternal chromatid, forming crossovers between the homologous chromosomes. The two homologous chromosomes separate along much of their length at diplotene, but they continue to be held together at localized chiasmata, which appear as X-shaped structures in micrographs. These physical links are thought to be the positions of crossing over. During metaphase and anaphase of the first meiotic division, the crossovers are gradually broken (with those at the ends resolved last) and the two homologous chromosomes (each still with two chromatids joined at a centromere) are moved into separate cells. During the second meiotic division (meiosis II), the centromere of each chromosome separates, allowing the two chromatids to move to separate cells, thus finishing the reductive division and making four haploid germ cells.

Meiotic Recombination

Mitotic recombination is a type of genetic recombination that may occur in somatic cells during their preparation for mitosis in both sexual and asexual organisms. In asexual organisms, the study of mitotic recombination is one way to understand genetic linkage because it is the only source of recombination within an individual. Additionally, mitotic recombination can result in the expression of recessive genes in an otherwise heterozygous individual. This expression has important implications for the study of tumorigenesis and lethal recessive genes. Mitotic homologous recombination occurs mainly between sister chromatids subsequent to replication (but prior to cell division). Inter-sister homologous recombination is ordinarily genetically silent. During mitosis the incidence of recombination between non-sister homologous chromatids is only about 1% of that between sister chromatids.

Occurrence

Mitotic recombination can happen at any locus but is observable in individuals that are heterozygous at a given locus. If a crossover event between non-sister chromatids affects that locus, then both homologous chromosomes will have one chromatid containing each genotype. The resulting phenotype of the daughter cells depends on how the chromosomes line up on the metaphase plate. If the chromatids containing different alleles line up on the same side of the plate, then the resulting daughter cells will appear heterozygous and be undetectable, despite the crossover event. However, if chromatids containing the same alleles line up on the same side, the daughter cells will be homozygous at that locus. This results in twin spotting, where one cell presents the homozygous recessive phenotype and the other cell has the homozygous wild type phenotype. If those daughter cells go on to replicate and divide, the twin spots will continue to grow and reflect the differential phenotype.

Mitotic recombination takes place during interphase. It has been suggested that recombination takes place during G1, when the DNA is in its 2-strand phase, and replicated during DNA synthesis. It is also possible to have the DNA break leading to mitotic recombination happen during G1, but for the repair to happen after replication.

Response to DNA Damage

In the budding yeast Saccharomyces cerevisiae, mutations in several genes needed for mitotic (and meiotic) recombination cause increased sensitivity to inactivation by radiation and genotoxic chemicals. For example, gene rad52 is required for mitotic recombination as well as meiotic recombination. Rad52 mutant yeast cells have increased sensitivity to killing by X-rays, methyl methanesulfonate and the DNA crosslinking agent 8-methoxypsoralen-plus-UV light, suggesting that mitotic recombinational repair is required for removal of the different DNA damages caused by these agents.

Mechanisms

The mechanisms behind mitotic recombination are similar to those behind meiotic recombination. These include sister chromatid exchange and mechanisms related to DNA double strand break repair by homologous recombination such as single-strand annealing, synthesis-dependent strand annealing (SDSA), and gene conversion through a double-Holliday Junction intermediate or SDSA. In addition, non-homologous mitotic recombination is a possibility and can often be attributed to non-homologous end joining.

Method

There are several theories on how mitotic crossover occurs. In the simple crossover model, the two homologous chromosomes overlap on or near a common Chromosomal fragile site (CFS). This leads to a double-strand break, which is then repaired using one of the two strands. This can lead to the two chromatids switching places. In another model, two overlapping sister chromosomes (there's no such thing as "sister chromosomes") form a double Holliday junction at a common repeat site and are later sheared in such a way that they switch places. In either model, the chromosomes are not guaranteed to trade evenly, or even to rejoin on opposite sides thus most patterns of cleavage do not result in any crossover event. Uneven trading introduces many of the deleterious effects of mitotic crossover.

Alternatively, a crossover can occur during DNA repair if, due to extensive damage, the homologous chromosome is chosen to be the template over the sister chromatid. This leads to gene synthesis since one copy of the allele is copied across from the homologous chromosome and then synthesized into the breach on the damaged chromosome. The net effect of this would be one heterozygous chromosome and one homozygous chromosome.

Advantages and Disadvantages

Mitotic crossover is known to occur in D. melanogaster, some asexually reproducing fungi and in normal human cells, where the event may allow normally recessive cancer-causing genes to be expressed and thus predispose the cell in which it occurs to the development of cancer. Alternately, a cell may become a homozygous mutant for a tumor-suppressing gene, leading to the same result. For example, Bloom's syndrome is caused by a mutation in RecQ helicase, which plays a role in DNA replication and repair. This mutation leads to high rates of mitotic recombination in mice, and this recombination rate is in turn responsible for causing tumor susceptibility in those mice. At the same time, mitotic recombination may be beneficial: it may play an important role in repairing double stranded breaks, and it may be beneficial to the organism if having homozygous dominant alleles is more functional than the heterozygous state. For use in experimentation with genomes in model organisms such as Drosophila melanogaster, mitotic recombination can be induced via X-ray and the FLP-FRT recombination system.

CHROMOSOMAL CROSSING OVER

Chromosomal crossover, or crossing over, is the exchange of genetic material between two homologous chromosomes non-sister chromatids that results in recombinant chromosomes during sexual reproduction. It is one of the final phases of genetic recombination, which occurs in the pachytene stage of prophase I of meiosis during a process called synapsis. Synapsis begins before the synaptonemal complex develops and is not completed until near the end of prophase I. Crossover usually occurs when matching regions on matching chromosomes break and then reconnect to the other chromosome.

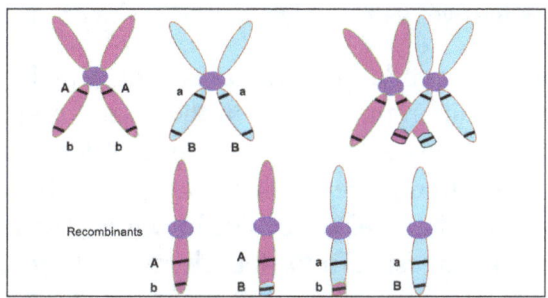

Crossing over occurs between prophase I and metaphase I and is the process where two homologous chromosome non-sister chromatids pair up with each other and exchange different segments of genetic material to form two recombinant chromosome sister chromatids. It can also happen during mitotic division, which may result in loss of heterozygosity. Crossing over is essential for the normal segregation of chromosomes during meiosis. Crossing over also accounts for genetic variation, because due to the swapping of genetic material during crossing over, the chromatids held together by the centromere are no longer identical. So, when the chromosomes go on to meiosis II and separate, some of the daughter cells receive daughter chromosomes with recombined alleles. Due to this genetic recombination, the offspring have a different set of alleles and genes than their parents do. In the diagram, genes B and b are crossed over with each other, making the resulting recombinants after meiosis Ab, AB, ab, and aB.

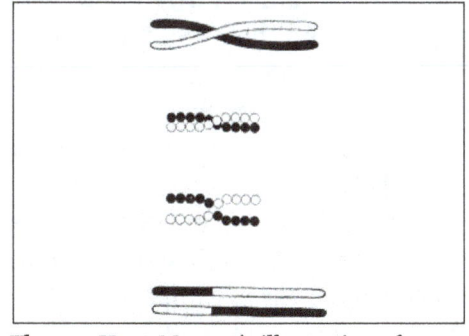

Thomas Hunt Morgan's illustration of crossing over.

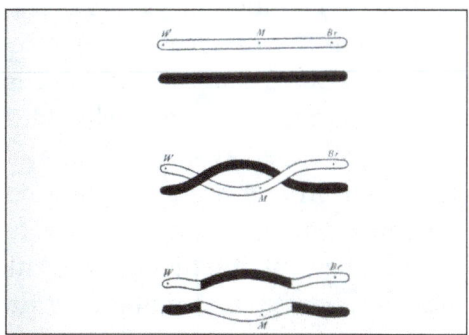

A double crossing over.

Crossing over was described, in theory, by Thomas Hunt Morgan. He relied on the discovery of Frans Alfons Janssens who described the phenomenon in 1909 and had called it "chiasmatypie". The term chiasma is linked, if not identical, to chromosomal crossover. Morgan immediately saw the great importance of Janssens' cytological interpretation of chiasmata to the experimental results of his research on the heredity of Drosophila. The physical basis of crossing over was first demonstrated by Harriet Creighton and Barbara McClintock in 1931.

The linked frequency of crossing over between two gene loci (markers) is the crossing-over value. For fixed set of genetic and environmental conditions, recombination in a particular region of a linkage structure (chromosome) tends to be constant and the same is then true for the crossing-over value which is used in the production of genetic maps.

There are two popular and overlapping theories that explain the origins of crossing-over, coming from the different theories on the origin of meiosis. The first theory rests upon the idea that meiosis evolved as another method of DNA repair, and thus crossing-over is a novel way to replace possibly damaged sections of DNA. The second theory comes from the idea that meiosis evolved from bacterial transformation, with the function of propagating diversity. In 1931, Barbara McClintock discovered a triploid maize plant. She made key findings regarding corn's karyotype, including the size and shape of the chromosomes. McClintock used the prophase and metaphase stages of mitosis to describe the morphology of corn's chromosomes, and later showed the first ever cytological demonstration of crossing over in meiosis. Working with student Harriet Creighton, McClintock also made significant contributions to the early understanding of codependency of linked genes.

DNA Repair Theory

Crossing over and DNA repair are very similar processes, which utilize many of the same protein complexes. In her report, "The Significance of Responses of the Genome to Challenge", McClintock studied corn to show how corn's genome would change itself to overcome threats to its survival. She used 450 self-pollinated plants that received from each parent a chromosome with a ruptured end. She used modified patterns of gene expression on different sectors of leaves of her corn plants show that transposable elements ("controlling elements") hide in the genome, and their mobility allows them to alter the action of genes at different loci. These elements can also restructure the genome, anywhere from a few nucleotides to whole segments of chromosome. Recombinases and primases lay a foundation of nucleotides along the DNA sequence. One such particular protein complex that is conserved between processes is RAD51, a well conserved recombinase protein that has been shown to be crucial in DNA repair as well as cross over. Several other genes in D. melanogaster have been linked as well to both processes, by showing that mutants at these specific loci cannot undergo DNA repair or crossing over. Such genes include mei-41, mei-9, hdm, spnA, and brca2. This large group of conserved genes between processes supports the theory of a close evolutionary relationship. Furthermore, DNA repair and crossover have been found to favor similar regions on chromosomes. In an experiment using radiation hybrid mapping on wheat's (Triticum aestivum L.) 3B chromosome, crossing over and DNA repair were found to occur predominantly in the same regions. Furthermore, crossing over has been correlated to occur in response to stressful, and likely DNA damaging, conditions.

Links to Bacterial Transformation

The process of bacterial transformation also shares many similarities with chromosomal cross over, particularly in the formation of overhangs on the sides of the broken DNA strand, allowing for the annealing of a new strand. Bacterial transformation itself has been linked to DNA repair many times. The second theory comes from the idea that meiosis evolved from bacterial transformation, with the function of propagating genetic diversity. Thus, this evidence suggests that it is a question of whether cross over is linked to DNA repair or bacterial transformation, as the two do not appear to be mutually exclusive. It is likely that crossing over may have evolved from bacterial transformation, which in turn developed from DNA repair, thus explaining the links between all three processes.

Chemistry

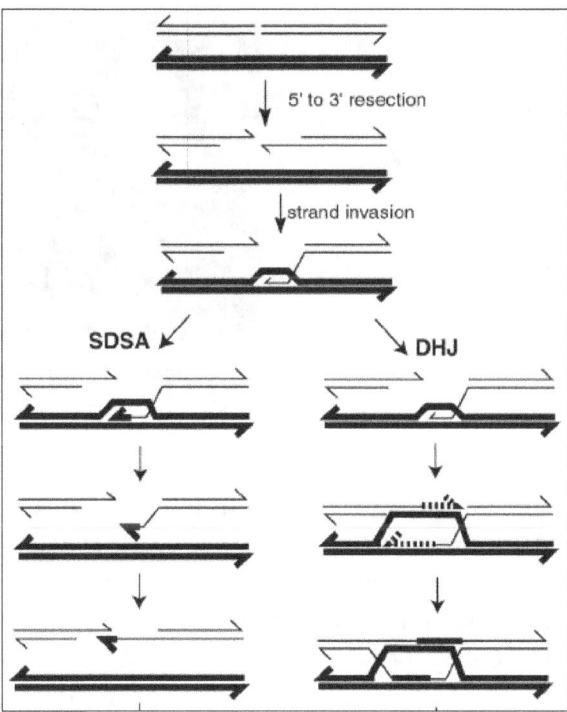

A current model of meiotic recombination, initiated by a double-strand break or gap, followed by pairing with a homologous chromosome and strand invasion to initiate the recombinational repair process. Repair of the gap can lead to crossover (CO) or non-crossover (NCO) of the flanking regions. CO recombination is thought to occur by the Double Holliday Junction (DHJ) model, illustrated on the right, above. NCO recombinants are thought to occur primarily by the Synthesis Dependent Strand Annealing (SDSA) model, illustrated on the left, above. Most recombination events appear to be the SDSA type.

Meiotic recombination may be initiated by double-stranded breaks that are introduced into the DNA by exposure to DNA damaging agents, or the Spo11 protein.[One or more exonucleases then digest the 5' ends generated by the double-stranded breaks to produce 3' single-stranded DNA tails. The meiosis-specific recombinase Dmc1 and the general recombinase Rad51 coat the single-stranded DNA to form nucleoprotein filaments. The recombinases catalyze invasion of the opposite chromatid by the single-stranded DNA from one end of the break. Next, the 3' end of the invading DNA primes DNA synthesis, causing displacement of the complementary strand, which subsequently anneals to the single-stranded DNA generated from the other end of the initial double-stranded break. The structure that results is a cross-strand exchange, also known as a Holliday junction. The contact between two chromatids that will soon undergo crossing-over is known as a chiasma. The Holliday junction is a tetrahedral structure which can be 'pulled' by other recombinases, moving it along the four-stranded structure.

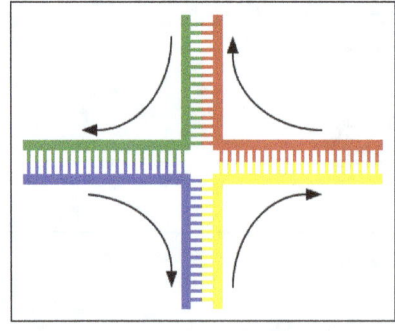

Holliday Junction.

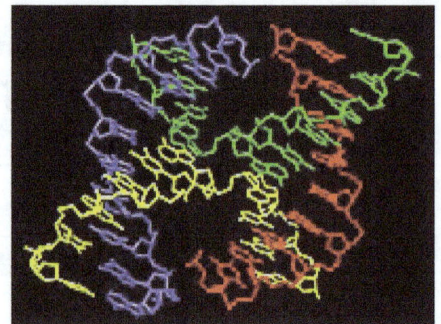

Molecular structure of a Holliday junction.

MSH4 and MSH5

The MSH4 and MSH5 proteins form a hetero-oligomeric structure (heterodimer) in yeast and humans. In the yeast Saccharomyces cerevisiae MSH4 and MSH5 act specifically to facilitate crossovers between homologous chromosomes during meiosis. The MSH4/MSH5 complex binds and stabilizes double Holliday junctions and promotes their resolution into crossover products. An MSH4 hypomorphic (partially functional) mutant of S. cerevisiae showed a 30% genome wide reduction in crossover numbers, and a large number of meioses with non exchange chromosomes. Nevertheless, this mutant gave rise to spore viability patterns suggesting that segregation of non-exchange chromosomes occurred efficiently. Thus in S. cerevisiae proper segregation apparently does not entirely depend on crossovers between homologous pairs.

Chiasma

The grasshopper Melanoplus femur-rubrum was exposed to an acute dose of X-rays during each individual stage of meiosis, and chiasma frequency was measured. Irradiation during the leptotene-zygotene stages of meiosis (that is, prior to the pachytene period in which crossover recombination occurs) was found to increase subsequent chiasma frequency. Similarly, in the grasshopper Chorthippus brunneus, exposure to X-irradiation during the zygotene-early pachytene stages caused a significant increase in mean cell chiasma frequency. Chiasma frequency was scored at the later diplotene-diakinesis stages of meiosis. These results suggest that X-rays induce DNA damages that are repaired by a crossover pathway leading to chiasma formation.

Consequences

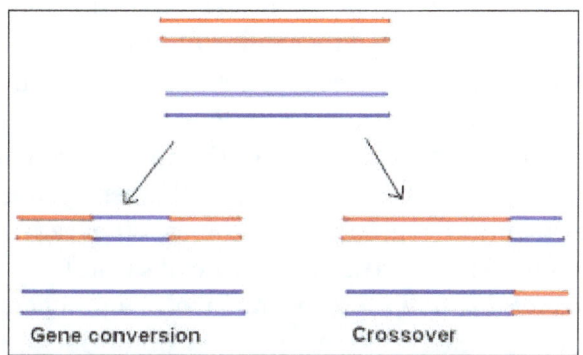

The difference between gene conversion and chromosomal crossover.

In most eukaryotes, a cell carries two versions of each gene, each referred to as an allele. Each parent passes on one allele to each offspring. An individual gamete inherits a complete haploid complement of alleles on chromosomes that are independently selected from each pair of chromatids lined up on the metaphase plate. Without recombination, all alleles for those genes linked together on the same chromosome would be inherited together. Meiotic recombination allows a more independent segregation between the two alleles that occupy the positions of single genes, as recombination shuffles the allele content between homologous chromosomes.

Recombination results in a new arrangement of maternal and paternal alleles on the same chromosome. Although the same genes appear in the same order, some alleles are different. In this way, it is theoretically possible to have any combination of parental alleles in an offspring, and the fact that two alleles appear together in one offspring does not have any influence on the statistical probability that another offspring will have the same combination. This principle of "independent assortment" of genes is fundamental to genetic inheritance. However, the frequency of recombination is actually not the same for all gene combinations. This leads to the notion of "genetic distance", which is a measure of recombination frequency averaged over a (suitably large) sample of pedigrees. Loosely speaking, one may say that this is because recombination is greatly influenced by the proximity of one gene to another. If two genes are located close to-gether on a chromosome, the likelihood that a recombination event will separate these two genes is less than if they were farther apart. Genetic linkage describes the tendency of genes to be inherited together as a result of their location on the same chromosome. Linkage disequilibrium describes a situation in which some combinations of genes or genetic markers occur more or less frequently in a population than would be expected from their distances apart. This concept is applied when searching for a gene that may cause a particular disease. This is done by comparing the occurrence of a specific DNA sequence with the appearance of a disease. When a high correlation between the two is found, it is likely that the appropriate gene sequence is really closer.

Non-homologous Crossover

Crossovers typically occur between homologous regions of matching chromosomes, but similarities in sequence and other factors can result in mismatched alignments. Most DNA is composed of base pair sequences repeated very large numbers of times. These repetitious segments, often referred to as satellites, are fairly homogenous among a species. During DNA replication, each strand of DNA is used as a template for the cre-ation of new strands using a partially-conserved mechanism; proper functioning of this process results in two identical, paired chromosomes, often called sisters. Sister chro-matid crossover events are known to occur at a rate of several crossover events per cell per division in eukaryotes. Most of these events involve an exchange of equal amounts of genetic information, but unequal exchanges may occur due to sequence mismatch. These are referred to by a variety of names, including non-homologous crossover, un-equal crossover, and unbalanced recombination, and result in an insertion or dele-tion of genetic information into the chromosome. While rare compared to homologous crossover events, these mutations are drastic, affecting many loci at the same time. They are considered the main driver behind the generation of gene duplications and are a general source of mutation within the genome.

The specific causes of non-homologous crossover events are unknown, but several in-fluential factors are known to increase the likelihood of an unequal crossover. One common vector leading to unbalanced recombination is the repair of double-strand breaks (DSBs). DSBs are often repaired using homology directed repair, a process which involves invasion of a template strand by the DSB strand. Nearby homologous regions of the template strand are often used for repair, which can give rise to either insertions or deletions in the genome if a non-homologous but complementary part of the template strand is used. Sequence similarity is a major player in crossover – cross-over events are more likely to occur in long regions of close identity on a gene. This means that any section of the genome with long sections of repetitive DNA is prone to crossover events.

The presence of transposable elements is another influential element of non-ho-mologous crossover. Repetitive regions of code characterize transposable elements;

complementary but non-homologous regions are ubiquitous within transposons. Because chromosomal regions composed of transposons have large quantities of identical, repetitious code in a condensed space, it is thought that transposon regions undergoing a crossover event are more prone to erroneous complementary match-up; that is to say, a section of a chromosome containing a lot of identical sequences, should it undergo a crossover event, is less certain to match up with a perfectly homologous section of complementary code and more prone to binding with a section of code on a slightly different part of the chromosome. This results in unbalanced recombination, as genetic information may be either inserted or deleted into the new chromosome, depending on where the recombination occurred.

While the motivating factors behind unequal recombination remain obscure, elements of the physical mechanism have been elucidated. Mismatch repair (MMR) proteins, for instance, are a well-known regulatory family of proteins, responsible for regulating mismatched sequences of DNA during replication and escape regulation. The operative goal of MMRs is the restoration of the parental genotype. One class of MMR in particular, MutSβ, is known to initiate the correction of insertion-deletion mismatches of up to 16 nucleotides. Little is known about the excision process in eukaryotes, but E. coli excisions involve the cleaving of a nick on either the 5' or 3' strand, after which DNA helicase and DNA polymerase III bind and generate single-stranded proteins, which are digested by exonucleases and attached to the strand by ligase. Multiple MMR pathways have been implicated in the maintenance of complex organism genome stability, and any of many possible malfunctions in the MMR pathway result in DNA editing and correction errors. Therefore, while it is not certain precisely what mechanisms lead to errors of non-homologous crossover, it is extremely likely that the MMR pathway is involved.

Mechanism and Significance of Crossing Over

Molecular Mechanism of Crossing Over

There are two important theories viz:

- Copy choice theory.

- Breakage and reunion theory to explain the mechanism of crossing over.

Copy Choice Theory

This theory was proposed by Belling. This theory states that the entire recombinant section or part arises from the newly synthesised section. The non-sister chromatids when come in close contact they copy some section of each other resulting in recombination. According to this theory, physical exchange of preformed chromatids does not take place.

The non-sister chromatids when come together during pairing, copy part of each other. Thus, recombinant chromosome or chromatids have some alleles of one chromatids and some of other. The information may be copied by one strand or both the strands. When only one strand copies, non-reciprocal recombinant is produced.

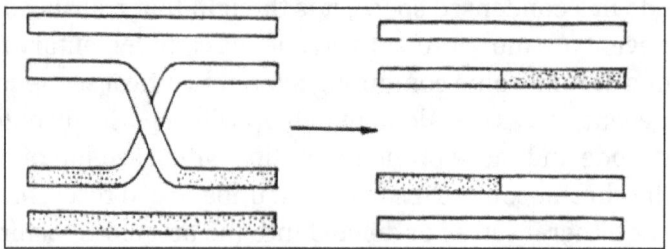

Crossing over according to copy choice theory.

If copy process involves both strands of chromosomes, reciprocal recombinants are produced. Assume, there are two chromosomes, viz., AB and ab. When their chromatids come in close contact they copy each other and result in Ab and aB re-combinations besides parental combinations.

This theory has two objections:

- According to this theory breakage and reunion does not occur, while it has been observed cytological.

- Generally crossing over takes place after DNA replication but here it takes place at the same time.

Breakage and Reunion Theory

This theory states that crossing over takes place due to breakage and reunion of non-sister chromatids. The two segments of parental chromosomes which are present in recombinants arise from physical breaks in the parental chromosomes with subsequent exchange of broken segments.

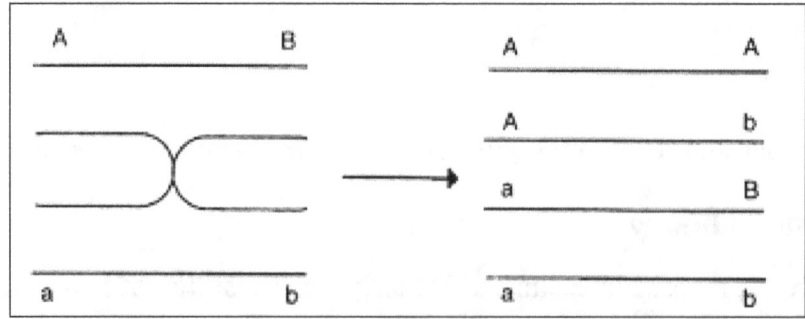

Crossing over according to breakage and reunion theory.

The breakage results due to mechanical strains that result from the separation of paired homologous chromosomes and chromatids in each chromosome during pachytene stage. The broken ends of non-sister chromatids unite to produce chiasmata resulting in crossing over.

Interference

The term interference was coined by Muller which refers to the tendency of one crossover to reduce the chance of another crossover in its adjacent region. Interference is affected by gene distance on the chromosome. Lesser the gene distance greater is the interference and vice versa. Generally, it is observed that crossing over in one region of chromosome may check the crossing over in the second region.

Sometimes, presence of recombination in one region enhances the chance of recombination in another adjacent region. This is termed as negative interference. This type of situation has been observed in some lower organisms, viz., Aspergillus and bacteriophages.

Coefficient of interference is estimated as follows:

Coefficient of interference (%) = 1 – Coefficient of coincidence x 100.

Positive and negative interference differ from one another in three main aspects.

Table: Differences between positive and negative interference.

Positive Interference	Negative Interference
One crossover reduces the chance of another crossover in the adjacent region.	One crossover enhance the chance of another crossover in the adjacent region.
Observed in both eukaryotes and prokaryotes.	Found in some lower organisms like Aspergillus and bacteriophages.
In this case coefficient of coincidence is less than one.	In this case coefficient of coincidence is always more than one.

Coincidence

This term was also coined by Muller to explain strength or degree of interference. The coefficient of coincidence is the percentage ratio of observed double crossovers to the expected double crossovers. The greater the coincidence, lesser will be the interference and vice versa. Thus,

$$\text{Coefficient of coincidence (\%)} = \frac{\text{Observed double crossovers}}{\text{Expected double crossovers}} \times 100$$

Coefficient of coincidence is a measure of the intensity of interference, because it has negative association with interference. The value of the coefficient of coincidence is less than 1 for positive interference, greater than 1 for negative interference, 1 for absence of interference and zero for complete or absolute interference.

Chromosome Mapping

Chromosome map refers to a line diagram which depicts various genes present on a chromosome and recombination frequency between them. Such maps are also known as genetic maps or linkage maps. The process of assigning genes on the chromosomes is known as chromosomal mapping.

The mapping of chromosomes is done with the help of three point test cross. A three point test cross is a cross of a trihybrid (F_1 differing in three genes) with its homozygous recessive parent.

The three point test cross provides useful information on two important aspects, viz:

- About the sequence of genes.

- About the recombination frequencies between genes. This information is essential for mapping of chromosomes.

Types of Crossing Over

Depending upon the number of chiasmata involved, crossing over may be of three types, viz., single, double and multiple as described below:

Single Crossing Over

It refers to formation of a single chiasma between non-sister chromatids of homologous chromosomes. Such cross over involves only two chromatids out of four.

Double Crossing Over

It refers to formation of two chiasmata between non-sister chromatids of homologous chromosomes. Double crossovers may involve either two strands or three or all the four strands. The ratio of recombinants and parental types under these three situations are observed as 2:2:3:1 and 4:0, respectively.

Multiple Crossing Over

Presence of more than two crossovers between non-sister chromatids of homologous chromosomes is referred to as multiple crossing over. Frequency of such type of crossing over is extremely low.

Factors affecting Crossing Over

- Distance: The distance between genes affects the frequency of crossing over. Greater the distance between genes higher is the chance of crossing over and vice versa.

- Age: Generally crossing over decreases with advancement in the age in the female Drosophila.

- Temperature: The rate of crossing over in Drosophila increases above and below the temperature of 22°C.

- Sex: The rate of crossing over also differs according to sex. There is lack of crossing over in Drosophila male and female silk moth.

- Nutrition: Presence of metallic ions like calcium and magnesium in the food caused reduction in recombination in Drosophila. However, removal of such chemicals from the diet increased the rate of crossing over.

- Chemicals: Treatment with mutagenic chemicals like alkylating agents was found to increase the frequency of crossing over in Drosophila female.

- Irradiation: Irradiation with X-rays and gamma rays was found to enhance the frequency of crossing over in Drosophila females.

- Structural Changes: Structural chromosomal changes especially inversions and translocations reduce the frequency of crossing over in the chromosomes where such changes are involved.

- Centromere Effect: Generally genes that are located adjacent to the centromere show reduced frequency of crossing over.

- Cytoplasmic Genes: In some species cytoplasmic genes also lead to reduction in crossing over. For example, Tifton male sterile cytoplasm in pearl millet.

Cytological Proof of Crossing Over

The first cytological evidence in support of genetic crossing over was provided by Curt Stern in 1931 on the basis of his experiments conducted with Drosophila. He used cytological markers in his studies. He selected a female fly in which one X-chromosome was broken into two segments.

Out of these two segments, one behaved as X-chromosome. The other X-chromosome had small portion of Y-chromosome attached to its one end. Thus, both the X-chromosomes in the female had distinct morphology and could be easily identified under microscope. In female fly, the broken X-chromosome had one mutant allele (carnation) for eye colour and another dominant allele (B) for bar eye shape.

The other X-chromosome with attached portion of Y chromosome had alleles for normal eye colour (red eye) and normal eye shape (oval eye). Thus, phenotype of female was barred. A cross of such females was made with carnation male (car+).

As a result of crossing over female flies produce four types of gametes, viz., two parental types or non-crossover types (car B and ++) and two recombinant types or crossover types (car+ and B+).

The male flies produce only two types of gametes (car + and Y), because crossing over does not take place in Drosophila male. A random union of two types of male gametes with four types of female gametes will produce males and females in equal number, means there will be four females and four males.

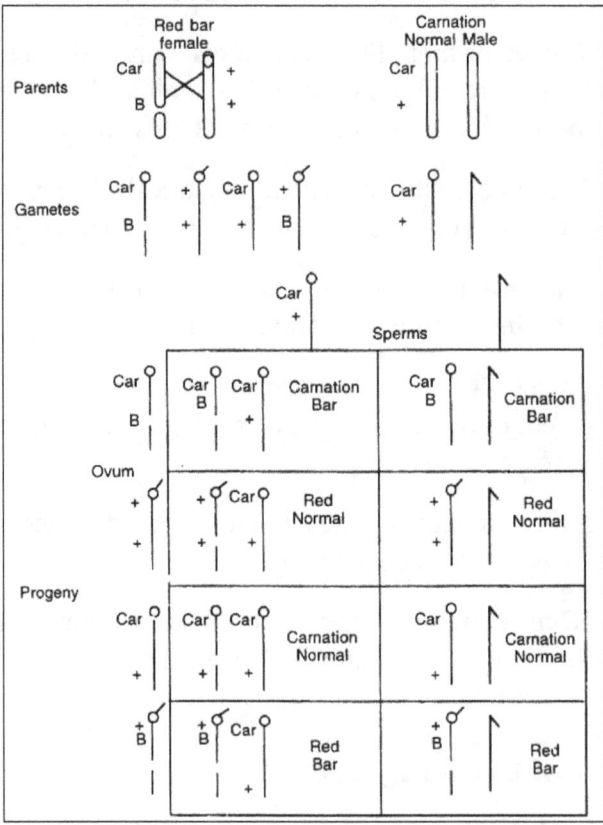

Cytological proof of crossing over in Drosphila.

Stern examined the chromosomes of recombinant types, viz., red bar and carnation normal under microscope. He observed that in carnation normal females both the X-chromosomes were of equal length. In red bar flies, one X-chromosome was normal and other was fragmented.

The fragmented X-chromosome also had attached part of Y-chromosome. Such chromosome combination in red bar is possible only through exchange of segments between non-sister chromatids of homologous chromosomes. This has proved that genetic crossing over is the result of cytological crossing over. Similar proof of cytological crossing over was provided by Creighton and McClintock in maize.

Significance of Crossing Over

Crossing over is useful in three principal ways, viz:

- Creation of variability.

- Locating genes on the chromosomes.

Preparing linkage maps as described below:

Creation of Variability

Crossing over leads to recombination or new combination and thus is a potential genetic mechanism for creating variability which is essential for improvement of genotypes through selection.

Locating Genes

Crossing over is a useful tool for locating genes in the chromosomes.

Linkage Maps

Crossing over plays an important role in the preparation of chromosome maps or linkage maps. It provides information about frequency of recombination's and sequence of genes which are required for preparation of linkage maps.

SOMATIC OR MITOTIC CROSSING OVER

In individuals heterozygous for a gene producing a visible phenotype, somatic crossing over leads to the production of "twin spots" of dominant and recessive phenotypes.

Mitotic crossing over was first discovered in Drosophila by Stern in 1936, between the six-linked genes y (yellow body colour) and sn (singed bristles = bent blunted bristles). The gene sn is located proximal, while y is located distal to the centromere of the X chromosome.

The phenotype of a fly in which both the genes are in heterozygous condition in repulsion phase (trans configuration) (sn^+ y/sn y^+) is wild type, so long there is no crossing over. There are three possibilities of crossing over.

- Crossing over between genes sn and y will produce cells with (sn^+ y/sn y) genotype and complementary combinations (,sn^+y^+/sn y^+). This would result into the production of a single, yellow spot (actually there is a twin spot of yellow and wild type but the latter is not detectable due to the wild type background).

- Crossing over between the centromere and gene sn would result in the twin combinations sn^+ y/sn^+ y and sny^+sn these would produce one spot with yellow body colour and the other having singed bristles.

- Double crossing over may occur between the centromere and gene sn and between the genes sn and y; this will produce the genotype sn y/sny$^+$. and the complementary combinations sn$^+$ y/sn$^+$ y$^+$. These genotypes will produce a single spot having singed bristles.

The size of twin spots varies depending on (he number of subsequent cell divisions in the crossover products. Mitotic recombination is less frequent than meiotic recombination. However, the relative frequencies for two different chromosomes or two regions of a chromosome may differ. In the fungus Aspergillus, mitotic recombination has been used in linkage studies and a linkage map was prepared for the genes pro (proline requiring mutation), paba (p-amino benzoate requiring mutation) and y (yellow mutation).

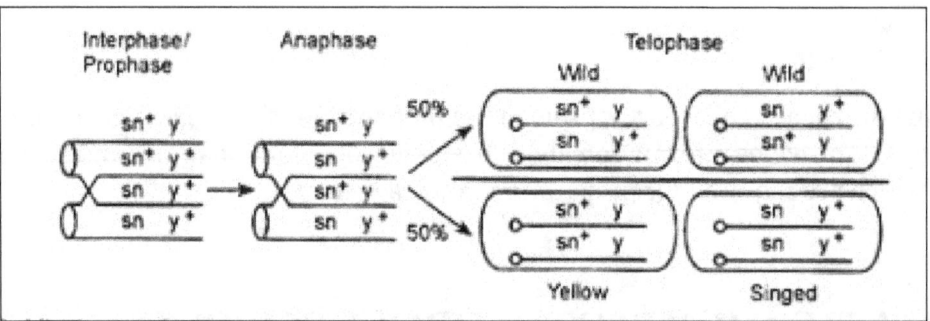

Mitotic crossing over in female Drosophila melanogaster X chromosome carrying singed (sn) and yellow (y) genes in heterozygous state. A single crossover between centromere and gene sn produces "yellow : singed" twin sectors.

It is presumed that mitotic crossing over occurs during mitotic prophase but the exact stage is not known. It is not known whether crossing over occurs during interphase before mitosis or during the transition from interphase to prophase or during the prophase.

GERMINAL OR MEIOTIC CROSSING OVER

Meiotic crossing over or germinal crossing over is a type of crossing over takes place in germinal cells during gametogenesis. This process is universal in occurrence and has great significance. Mechanism of meiotic crossing over.

Crossing over is a crucial process that generates genetic difference within a population. The prerequisites for crossing over are firstly, 99.7% of DNA replication and 75% of histone synthesis must occur by prophase I. Secondly, each chromosome must attach by its telomeres (ends of the chromosome) to the nuclear envelope through specialized structures called attachment plaques.

The major steps in meiotic crossing over are:

- Synapsis,

- Duplication of chromosome,

- Crossing over,

- Terminalisation.

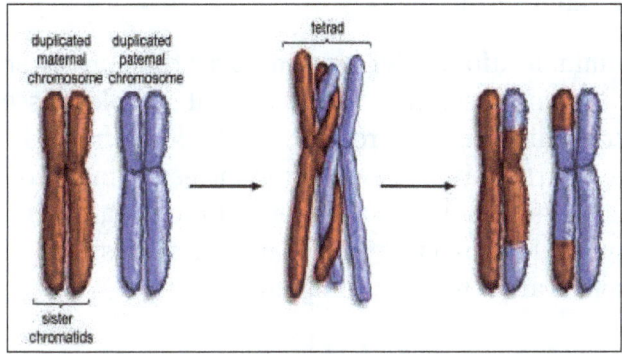

Synapsis is the intimate pairing between the two homologous chromosomes. It is initiated during the zygotene stages of prophase I of meiosis I. Here chromosomes are aligned side by side and each gene has its counterpart aligned perfectly (gene for gene alignment). The resultant pairs of homologous chromosomes are called bi-valents.

Synapsis is followed by duplication of chromosome (in pachytene). Sister chromatids are held at centromere. At this stage, each bi-valent has four chromatids now called as tetrad.

Crossing over or exchange of segments between the non-sister chromatids of homologous chromosome occurs at the tetrad stage. Homologous chromosome may stay in synapsis for even days during pachytene stage. Now let us have a look into the details of crossing over. For easier understanding, crossing over can be divided into three major steps:

- Breakage of chromatid segments.

- Their transposition (movement to the respective site).

- Fusion or joining.

Enzymes Involved in Crossing Over

- Recombinase is the major enzyme regulating recombination event.

- Endonuclease.

- Ligase enzyme.

Endonuclease is responsible for breakage of 2 non-sister chromatids at corresponding sites. This is followed by the exchange of segments and finally the exchanged segments are joined or the gap is filled by ligase enzyme.

Crossing over takes place at several points on a tetrad and result in several chiasmata. These are regions were chromosomes are held together. Larger the chromosome size the more the number of chiasmata. Frequency of crossing over is dependent on the physical distance between genes on the chromosome. The chance of crossing over is more for distantly located genes.

The final step is terminalisation. After crossing over the non-sister chromatids starts to repel each other. During diplotene, Synaptonemal complex dissolves and desynapsis takes place. During diakinesis, chromosome detaches from the nuclear envelope and the chromatids separate progressively from the centromere towards the chiasmata. Meanwhile chiasma itself moves in a zipper fashion towards the end of tetrad. This movement of chiasma is known as terminalisation. As a result of terminalisation, homologous chromosomes are separated completely.

References

- Alberts B, Johnson A, Lewis J, Raff M, Roberts K, Walter P (2008). Molecular Biology of the Cell (5th ed.). Garland Science. p. 305. ISBN 978-0-8153-4105-5

- Recombination-genetics, science: britannica.com, Retrieved 9 May, 2020

- A-Advantages-of-Genetic-Recombination, A-Recombination-of-DNA, A-Replication-Maintenance-and-Alteration-of-the-Genetic-Material, A-Working-with-Molecular-Genetics-(Hardison), Genetics: bio.libretexts.org, Retrieved 22 April, 2020

- Bernstein, H; Bernstein, C (2010). "Evolutionary origin of recombination during meiosis". BioScience. 60 (7): 498–505. doi:10.1525/bio.2010.60.7.5

- Crossing-over-process-behind-our-2012: biologyexams4u.com, Retrieved 17 August, 2020

- Crossing-over-meaning-mechanism-and-significance-genetics-37840, crossing-over, genetics: biologydiscussion.com, Retrieved 17 June, 2020

- Lee, Phoebe S.; Greenwell, Patricia W.; Dominska, Margaret; Gawel, Malgorzata; Hamilton, Monica; Petes, Thomas D. (2009). "A Fine-Structure Map of Spontaneous Mitotic Crossovers in the Yeast Saccharomyces cerevisiae". PLoS Genet. 5 (3): e1000410. doi:10.1371/journal.pgen.1000410. PMC 2646836. PMID 19282969

- Mitotic-crossing-over-cell-division-36171, cell-division-cell, cell: biologydiscussion.com, Retrieved 11 April, 2020

Permissions

Index

A

Abscission, 137, 140

Allele, 17, 46, 206, 210, 215, 221

Aneuploidy, 4, 19, 24, 28, 39, 107, 110, 114, 124, 128-129

Anti-mitogen Signaling, 106

Apoptosis, 5-6, 53-54, 70-71, 74, 88-103, 119, 129, 149

Apoptotic Cell Disassembly, 95

Aurora Kinase Inhibitors, 3-4

B

Bdelloid Rotifers, 55

Binary Fission, 2, 104, 108, 133-134, 136-137

Biochemical Regulators, 78

Biomarkers, 4

Budding, 2, 32-33, 40, 43, 49, 51, 115, 133, 165, 195, 197, 199, 209

C

Caspase-independent Apoptotic Pathway, 94

Caspases, 89, 91, 93-94, 96-97, 99-100, 102-103

Cell Cycle, 1, 4-6, 9, 12, 16, 39, 53, 56, 65, 82-84, 86, 99-101, 107-108, 111, 114, 124, 129, 132, 137, 140, 142, 164-165, 171, 181, 184, 194-195

Cell Differentiation, 65, 132, 141-142, 145-146, 150-156

Cell Plate, 137, 140-141

Chiasma, 40, 174, 212, 214-215, 220, 226

Chromosome Mapping, 219

Cohesin, 24-29, 31-41, 111-113, 121, 127, 130, 163-164

Cohesion, 19, 24-29, 32-33, 35, 37-41, 109, 111-114, 118, 121, 130

Crossover, 19, 23, 39, 46, 48, 174, 193-194, 197-198, 206, 209-217, 219, 221, 224

Cytokinesis, 3, 7-8, 57, 137-141, 147, 158, 164, 167-169, 172, 177, 184, 186

D

Dedifferentiation, 144-145

Differentiated Bone, 74

Differentiated Muscle, 73, 144

Dna Methylation, 148-149

Dna Repair Theory, 212

Dna Replication, 6-7, 11-12, 15, 24, 29, 50, 52-54, 59, 79-81, 83-84, 104, 108-109, 112, 127, 130, 133-134, 170-172, 194-195, 199, 210, 216, 218, 224

E

Epigenetic Markers, 71

Extracellular Signals, 104

Extrinsic Pathway, 89-91, 102

F

Fas Path, 93

G

Genotype, 209, 217, 223-224

H

Heterochromatin, 26, 28, 112, 122, 129, 149

Histone Acetylation, 148-149

Histone Methylation, 71, 149

Histone Synthesis, 81, 224

Homologous Chromosome, 7, 33, 48, 56, 58, 174, 196-197, 207-208, 210-211, 213, 225

Homologous Recombinational Repair, 44, 84, 175

Hyperactive Apoptosis, 100

I

Interphase, 59, 62, 69, 77-78, 83, 85-86, 99, 117, 130, 137, 140, 158-159, 164, 166-167, 169, 177-178, 188, 192, 209, 224

Intrinsic Pathway, 89-90, 102

K

Kinesin Spindle Protein Inhibitors, 5

Kinetochore, 13, 18, 110-112, 114-131, 138, 140, 158, 160-163, 175-176, 179, 181-182, 184

M

Matrix Elasticity, 151

Mature Hepatocytes, 73

Mitogen Signaling, 105-106

Mitotic Inhibitors, 3, 5

N
Necrotic Cells, 97
Noncrossover, 48
Nucleosome Replication, 81
Nutrient Signaling Pathways, 75

O
Outcrossing, 51-52, 54-55

P
Phenotype, 77, 83, 96, 138, 148, 209, 221, 223
Polo-like Kinase Inhibitors, 4
Polycomb Repressive Complex, 82, 148
Polyploidy, 12-13
Progeny, 2, 14, 22-23, 51, 54, 71, 98, 133
Propel Chromosome Movement, 121, 130

Q
Quiescent Stem Cells, 71

R
Recombination, 2, 7-16, 18-24, 33, 38-53, 55-59, 83, 135, 173-177, 184, 190-217, 219-221, 223-225
Restriction Point, 70, 79-80, 104-106, 119

S
Segregation, 2-3, 9-13, 18-19, 22, 24, 26-29, 33, 39-40, 46, 57, 59, 96, 104, 107-110, 116, 120-121, 125, 127, 131, 153, 163, 174, 211, 214-215
Senescent Cells, 71, 73
Somatic Cells, 28-29, 38, 44, 46, 53-54, 77-78, 134, 143, 145, 160, 183, 194, 208
Spatial Regulation, 83, 87
Spindle Assembly, 3-4, 84, 108, 111, 114-115, 117, 124, 129, 158, 165-167
Spindle Checkpoint Activation, 129-130
Switch-like Activation, 85
Synapsis, 9-10, 16, 26, 30, 32-38, 40, 49-50, 52, 174, 207-208, 211, 225
Syngamy, 9, 15-16, 42

T
Telophase, 7-9, 57, 59, 129, 137, 157-158, 163-169, 172-173, 182, 184-186, 207
Terminalisation, 225-226
Transcriptomes, 71

Z
Zygote, 2, 133, 141, 143, 152, 184, 207

www.ingramcontent.com/pod-product-compliance
Lightning Source LLC
Chambersburg PA
CBHW080402190526
45161CB00003B/106